P9-AQA-714

Contextual Analysis: Concepts and Statistical Techniques

Lawrence H. Boyd, Jr.
University of California, Berkeley

Gudmund R. Iversen
Swarthmore College

Wadsworth Publishing Company
Belmont, California
A division of Wadsworth, Inc.

Sociology Editor: Curt Peoples
Production: Cobb/Dunlop Publisher Services, Inc.
Designer: Marsha Cohen
Technical Illustrator: SoHo Studios, Inc.

Printed in the United States of America

1 2 3 4 5 6 7 8 9 10–82 81 80 79

Library of Congress Cataloging in Publication Data

Boyd, Lawrence H
Contextual analysis.

Bibliography: p.
Includes index.
1. Social sciences—Statistical methods. 2. Social sciences—Methodology. I. Iversen, Gudmund R., joint author. II. Title.
H61.B6368 300'.1'82 79-16958
ISBN 0-534-00693-0

Foreword

Some forty years ago, when I first came to think of myself as a Social Scientist, with capital S's, there were two words that I found tantalizing, titillating, and bamboozling because they were, on the one hand, intuitively intriguing but, on the other hand, altogether untractable for purposes of empirical inquiry. One of them was the word "configurational," the other was the word "contextual." I think I came across them first in the early work of the political scientist Harold D. Lasswell, but possibly also in the work of the social psychologist Kurt Lewin.

What made the conception of "individual-in-context" or "unit-in-environment" so intriguing was the fact that it reflected the intellectual tension between the quest to study human phenomena as "wholes" that the social sciences had inherited from the biological sciences, and the quest to reduce wholes to their constituent parts through what was then known as operationalism. In fact, however, not much happened at the frontiers of empirical social research until after the war when, at Columbia University, the sociologists Paul F. Lazarsfeld, Robert K. Merton, and their students began to treat the person not as an isolated individual but as a social being. Their work was the starting point for the kind of modern contextual analysis now explicated in this sophisticated statistical volume by Boyd and Iversen.

There was more trouble in understanding and using the word "configurational." I knew that it had something to do with the "structure of fields," but it was not until I had become more familiar with Lewin's notion of "life space" that I could make some coherent sense of it. The life spaces of many individuals were said to constitute interlocking fields in which persons are related to each other in complex ways. But, for reasons much too complicated to mention here, Lewinian field theory, whatever its stimulating qualities, did not take off in the social sciences and, as far as I know, is languishing in social psychology.

Nevertheless, the notion that the individual's context is somehow structured or patterned to make for identifiable configurations rather than being simply aggregative did take root in the social sciences in the fifties, especially in the work of some English social anthropologists and notably in the work of S. F. Nadel. The term then coming into prominence was "network" which, it seems, has more direct and concrete meaning than "configuration." Nadel used it to discover linkages between and among social roles. Of course, such terms as network, web, or mesh had long been common as suggestive metaphors in the language of the social sciences, but they had to be given more than literary definitions before being scientifically useful. Although mathematical graph theory and other metrics are now available, the study of configurations or networks of social relations yet awaits its first systematic exposition as that provided here for contextual analysis by Boyd and Iversen.

What makes such systematic exposition of configurational analysis in terms of statistics all the more necessary and desirable are the very difficulties and dilemmas of contextual analysis to which Boyd and Iversen address themselves throughout the book, but especially in the last chapter (which should be comprehensible even to scholars who are not too familiar with the language of multiple regression). The analysis of social contexts through multilevel treatment of individual and grouped variables as well as their (statistical) interactions is an enormous step forward in the social sciences; yet it remains essentially aggregative in the sense that it discovers mathematical rather than real social structures. That the individuals or groups located in social contexts actually interact with and behave in structured ways toward each other is an essential assumption of contextual analysis, but it is an assumption that cannot be proved as such by contextual analysis. To do so, what is clearly needed is configurational or network analysis that traces social relationships from one person to the next, from person to group, from group to group, and so on. This is not to say that contextual analysis and configurational analysis are at loggerheads; on the contrary, many of the statistical cautions which Boyd and Iversen so correctly entreat us to observe could be resolved if one knew in fact the direction of the causal flows in social interaction which, in contextual analysis, must remain matters of theoretical speculation. I therefore see multilevel contextual analysis and configurational analysis as two parts of the methodological package that will ultimately help us to close the "micro–macro gap."

As one reads this work on the statistics of multilevel analysis by Boyd and Iversen, the growing literature to which they refer, and some recent work on network analysis by Harrison C. White and others, what strikes one immediately is how far both theory and method have outdistanced empirical investigations. Boyd and Iversen are able to draw only on surprisingly little empirical material that does justice to their exposition of the statistical possibilities. They are of course fully aware of this poverty in data, just as are those who would now want to pursue configurational analysis. Hopefully, therefore, the recent theoretical and methodological developments in both contextual and configurational analysis will inspire researchers to formulate their study designs in ways that permit the collection of relevant data at several levels and in several settings of social complexity.

I cannot but emphasize the importance for empirical social science of the kind of analysis presented in this volume by Boyd and Iversen. The analysis itself as now conducted would have been impossible prior to the digital computer revolution beginning in the early sixties. If one recalls the controversies over various fallacies of inference that remained so unresolved in the fifties, or the frustrated efforts of that period to extend small-group sociometry to large collectivities, one is struck by the fact of how much the modern computer technology has transformed the management of massive data sets and their statistical analysis, both of survey and aggregate data. Just as early contextual analysis by way of contingency tables has now been replaced by the kind of multilevel treatment through multiple regression described by Boyd and Iversen, so configurational or network analysis can now deal with large and complex data sets in order to trace actual social relations and interactions. But failure to invest adequate financial resources in the collection of data relevant to multilevel and configurational investigations remains a very real obstacle in the use of the new technology and methodology. Despite demonstrated capacity, foundations and governments remain reluctant or unwilling to support the social sciences at a level of expenditure that is taken for granted in the natural sciences.

Contextual, multilevel analysis has come to constitute a stream that has many tributaries. I have mentioned a political scientist, a social psychologist, a social anthropologist, and two sociologists as those who influenced my own thinking in this matter. The point I want to make is, of course, that thinking in terms of the relationship between micro and macro phenomena is generic to all the social sciences and that, therefore, this volume on multilevel analysis deserves the attention of scholars across all the social sciences. Provided students have some rudimentary knowledge of multiple regression, the book should prove useful in introductory courses in social science methodology, for it presents basically an elaboration of a single, easy-to-grasp statistical model. I used an early draft of the book in one of my courses, working through it with my students for several sessions. It is an experience very much as one has when following and traversing the canals of Venice: there are many small canals to travel and many bridges to cross, but one always comes back to the Canal Grande.

HEINZ EULAU
William B. Munro Professor
of Political Science
Stanford University

Preface

The Problem

To what extent are we affected by characteristics of the groups we belong to, and to what extent are we affected by our individual characteristics? This question lies at the very center of social science inquiry. The literature in this area is rich and varied, but the main emphasis has been more along substantive lines rather than on how to actually study the possible presence of individual and group effects and how to measure their magnitudes. In this book we address ourselves to these more methodological issues.

We begin by noting that when two variables are not related in the same way for individuals in different groups, there must be something about those groups themselves that affects the relationship. We assume that the regression coefficients within the groups depend on certain group variables, and with such a model we study the effects of both individual and group variables. These effects are measured by the magnitudes and sizes of a series of regression coefficients as well as various decompositions of sums of squares.

This model goes beyond other ways of looking at group effects such as the analysis of variance, analysis of covariance, or contingency-table analyses of various kinds.

The Audience

Reading this book requires a working knowledge of multiple-regression analysis on a level usually achieved after a typical sequence of two courses in statistics for social science students. This makes the book suitable for advanced methods or statistics courses. We also think the book can be read with interest by social scientists who have substantive interests in contextual analysis as well as by those with methodological interests in this area.

Many of the computations involved in the examples given in this book can be performed with the assistance of a pocket calculator. However, computer assistance is frequently required. To assist the inexperienced computer user and to supplement the formal discussion of multilevel methods, a number of illustrations of computer applications are presented in Appendix A. Because of general availability and common use, the *Statistical Package for Social Sciences* (SPSS) is used to illustrate the procedures.

The Organization of **Contextual Analysis**

The book is formally organized around five parts, with a common statistical model running through the entire book. Parts II and III represent the core of the book. Two complementary views of the basic model are explored there. Part II focuses on the relationship between variables for a set of individuals and asks what role group variables play in determining that relationship. Part III begins with the relationship between variables for a set of groups and asks what role individual-level variables play.

Part I introduces the basic model in terms of the objectives of multilevel analysis and alternative approaches. Since the two chapters in Part I proceed directly to the basic model of contextual analysis, readers who are uninitiated in contextual analysis should profit by rereading these chapters after working through Parts II and III.

For the study of individual and group variables, a basic requirement is that data be available both for the individuals in the study and for the groups to which they belong. These data requirements often impose formidable obstacles to a complete analysis. For this reason, the whole of Part IV is devoted to methods that can be used with less than complete data or when data come from different sources. Because the ways such opportunities can occur are too numerous to list and treat separately, we discuss a number of cases we think occur most frequently.

Early efforts at contextual analysis raised many questions about their adequacy. One criticism was that they oversimplified reality by considering only one set of explanatory variables or by considering only one level of group effects. Other criticisms were aimed at the definition and measurement of group variables. These and other issues in multilevel analysis are identified and discussed in Part V. That discussion points out how the contents of this book deal with many of the shortcomings of earlier work.

Special Issues

There are many statistical equations in this book, but the book is not meant to be a textbook in statistics. We try to keep the formalities, including the notation, to a minimum without losing precision.

For example, when we say that we study the relationship between two variables according to the equation

$$\begin{aligned} y_i &= a + bx_i + e_i \\ &= 5.0 + 10.0x_i + i \end{aligned}$$

we have assumed an underlying linear regression model with parameters α and β and residuals ϵ_1, ϵ_2, $\cdots$, ϵ_n. By using ordinary least squares the coefficients a and b become the estimators of the corresponding parameters α and β. The particular numerical estimates in this example are 5.0 for α and 10.0 for β.

Occasionally we refer to other estimating procedures, particularly some using weighted regression. In those cases the coefficients in the equations are the estimators resulting from such procedures. Certain other estimating procedures using incomplete data are discussed in Part IV. It should be clear from the context what procedure is used to arrive at any particular estimator.

We do not do much with statistical inference in this book. The standard formulas for t and F tests are based on simple random samples. Given the unique and generally severe data requirements for contextual analyses, it is difficult to imagine an analysis being done on data collected according to a plan that simple. Furthermore, unless the data sets are small, most results are statistically significant anyway, and it is more useful to know how large the various effects are than that they are statistically significant.

A number of important methodological insights and technical considerations emerge in the discussion of proposed strategies and procedures. For example, there are conditions under which alternative approaches to the same data produce different results. Diagnostically, this can be taken as an indication that the model under consideration is not properly specified. This material is contained in the Appendixes of the book. The Appendixes also contain procedures which open the door to the contextual analysis of large stores of data in table form, such as the census.

The Authors

We came to these issues and problems by different paths. Boyd started with the role of groups in the theory and practices of social work and combined that with his interest in political science and social science methodology. Iversen had worked on some of the statistical problems that arise when dealing with individual- and group-level data, as part of his general interest in the use of statistics in the social sciences. Our paths merged and we began collaborating at the Institute for Social Research at The University of Michigan. After leaving there the work continued on the two coasts, aided by short trips, the phone, and the mail.

Acknowledgments

We are grateful for the support we received from the National Science Foundation capability-building grant (GI-29904) to the Institute for Social Research, The University of Michigan, from the Faculty Research Support Committee at Swarthmore College, and from the Committee on Research at the University of California, Berkeley.

We wish to extend our thanks to Donald E. Stokes whose fascination with the problems of data on individuals and groups was so contagious as to lead us in the direction of this book. Lutz Erbring gave us the benefit of a penetrating review of an earlier draft. Heinz Eulau's enthusiasm for our work gave us a welcome boost at a critical time. We are also grateful to Glen Firebaugh, Vanderbilt University, and Jerry Medler, University of Oregon, whose thoughtful suggestions improved the manuscript in numerous ways.

We wish to express personally our appreciation to Professor Milton Chernin, recently retired Dean of the Berkeley School of Social Welfare, and all the doctoral students of that school for their friendship and encouragement.

We were blessed with superbly competent and devoted research assistants. Our very special thanks go to Linda Remy, Wes Boyd, and Marlene Clark for their superior editing and technical assistance, and for their extra special hard work and dedication to this project.

The shortcomings that remain can only be blamed on us.

Lawrence H. Boyd, Jr.

Gudmund R. Iversen

Berkeley and Swarthmore

Contents

List of Tables and Figures

Chapter 1

Tables

Figures

Chapter 2

Tables

No figures in Chapter 2

Chapter 3

Tables

Figures

Chapter 4

Tables

Figures

Chapter 5

Tables

Figures

Chapter 6

Tables

Figures

Chapter 7

Tables

Figures

Chapter 8

Tables

Figures

Chapter 9

Tables

Figures

Chapter 10

Tables

No figures in Chapter 10

Chapter 11

Tables

No figures in Chapter 11

Chapter 12

Tables

Figures

Chapter 13

No tables or figures

I

The Objectives of Multilevel Analysis

Social research analyses typically involve individuals *or* groups. Multilevel analyses involve both. The two chapters in Part I introduce multilevel analysis by discussing its two major objectives. This discussion immediately involves the reader in the logic and procedures of multilevel analysis. Its purpose is to preview and organize the materials in the rest of the book and to provide a framework for identifying the important literature.

To avoid complications identified in later discussion, points are illustrated with hypothetical data. Actual examples and bibliographical references to relevant studies illustrate potential applications and provide an historical perspective on multilevel methodology. Descriptions of computer operations for each example are contained in Appendix A, Illustrations 1 through 7.

Discussion in this part proceeds directly to formal linear models of contextual analysis. Readers who are relatively inexperienced with general linear multiple regression or are completely unfamiliar with contextual analysis will profit from a rereading of Part I after having worked through the materials in Part II.

Introduction to the First Objective of Multilevel Analysis: Explaining Variation in Individual Behavior

The first of the two broad objectives of multilevel analysis examined in this book is to explain variation in individual-level behaviors in terms of individual- and group-level effects. The purpose of this chapter is to locate the basic model of contextual analysis considered throughout this book in a methodological framework based on this specific objective, rather than an historical framework dealing only with the contextual model. Other approaches, including those based on analysis of variance, analysis of covariance, and various contingency-table approaches, are described. This is followed by an introduction to the basic model of contextual analysis. For comparison the various methods are illustrated by a single hypothetical data set. The presentation is intended to be methodologically integrative. However, the history of the development can be gathered from the references.[1]

Part II of this book is devoted to a step-by-step examination of the contextual model.

[1] The literature involving multilevel analysis is rich and varied. The References at the end of this book cover a wide range of this literature. The following is a list of works more directly related to the discussion in this chapter: Alexander and Eckland (1975); Alwin (1976); Bachman et al. (1966); Blau (1960); Bowers (1968); Boyle (1966a, 1966b); Campbell and Alexander (1965); Cohen (1968); Davis (1961); Davis et al. (1961); Ennis (1962); Firebaugh (1977); Flinn (1970); Hanushek et al. (1974); Hauser (1971); Kish (1962); Kleinbaum and Kupper (1978); Lazarsfeld and Menzel (1961); McDill et al. (1967); Meyer (1970); Michael (1961, 1966); Miller and Erickson (1974); Namboodiri et al. (1975); Nelson (1972); Przeworski (1974); Przeworski and Teune (1970); Putnam (1966); Rigsby and McDill (1972); Rosenberg (1962); Schuessler (1969); Sewell (1964); Sewell and Armer (1966); Tannenbaum and Bachman (1964); Tannenbaum and Smith (1964); Turner (1966); Werts and Linn (1971); Wilson (1959).

Other pertinent works address more directly the methodological issues in contextual analysis. These works are cited in the discussion of issues in Chapter 13. There also are many works which are indirectly relevant to contextual analysis. These focus more directly on other problems in multilevel analysis, such as the aggregation problem and the problem of cross-level inference. References to these works are contained in Chapter 2.

Computer operations for this chapter are described in Appendix A, Illustrations 1 through 6.

1.1

Explaining Variation in an Individual-level Variable with an Individual-level Variable in a Single Subgroup

Consider a hypothetical study of attitudes toward liberalizing abortion laws in a particular state. A measure of opposition is based on a scale ranging from 0 to 100, where the high values indicate opposition. Data from the researcher's community are presented in Table 1.1. The number of observations is purposely small so that the subsequent discussion of concepts and procedures can easily be connected to the data. Computer operations are shown in Appendix A, Illustration 1.

Table 1.1
Hypothetical Observations on Opposition Scores and Religiosity in a Single Subgroup

OPPOSITION TO LIBERALIZED ABORTION LAWS		RELIGIOSITY	
Y	$\bar{Y}$	X	$\bar{X}$
0	20	0	3
10		0	
15		0	
25		2	
5		3	
45		3	
15		4	
10		6	
30		6	
45		6	

The average score of 20 indicates generally low opposition to liberal reform. However, the individual scores range from 0 to 45, indicating that individuals vary in their opposition to liberalizing abortion laws. The researcher hopes to explain this variation.

One theory is that differences between individuals on religious commitment explain at least part of the variation in abortion attitudes. The variable "religiosity" is based on a scale ranging from 0 to 10, where the high values indicate high religious commitment. Observations on this variable are shown in Table 1.1. The dependent variable is denoted by the symbol Y and the explanatory variable by X.

A conventional way to study the relationship between Y and X is with a simple regression model as in

$$y_{ic} = a_c + b_c x_{ic} + e_{ic}, \qquad i = 1, 2, \ldots, n \tag{1.1}$$

In Eq. (1.1) the symbols a_c and b_c represent the regression intercept and slope. The subscript c is used to indicate that the results apply to a particular community. The residual term e_{ic} represents all unspecified effects on the dependent variable Y. Examples could include socioeconomic status, religious affiliation, education, age, and sex. Estimates for the intercept a_c and the slope b_c are

$$y_{ic} = 10.9 + 3.0x_{ic} + e_{ic}, \qquad R^2 = 0.229 \tag{1.2}$$

The numerical value of the intercept in Eq. (1.2) indicates that a resident of the selected community with a score of 0 on religiosity is predicted, on the average, to have a score of 10.9 on opposition to liberalized abortion laws. The slope indicates that for each unit increase in religiosity, an increase of 3.0 units in opposition is expected. The value of R^2 indicates that 22.9 percent of the variation in abortion-law opposition is explained by religiosity. Assuming a more typical sample size, these results suggest that religiosity is a significant predictor of abortion attitudes.

1.2

Explaining Variation in an Individual-level Variable with an Individual-level Variable in a General Population

But the particular community surveyed is not representative of the general (state) population. Notably, community averages on abortion attitudes and religiosity are both lower than expected. The study is consequently broadened to include a sample of communities. For each community the researcher draws a random sample of individuals. This design allows a study of the implications of community differences in the analysis of abortion attitudes.

The data for the selected communities are shown in Table 1.2. These observations represent a more general population (the state) of which the first sample was but a part. The relationship between Y and X for all the observations in Table 1.2 taken together is analyzed according to the equation

$$y_{is} = a_s + b_s x_{is} + e_{is}, \qquad i = 1, 2, \ldots, n_s \tag{1.3}$$

In Eq. (1.3) the subscript s shows that this equation applies to the entire state as opposed to a subgroup (community) analysis. Computer procedures for this analysis are presented in Appendix A, Illustration 2.

Table 1.2
Hypothetical Observations for the General (State) Population

COMMUNITY	OPPOSITION TO LIBERALIZED ABORTION LAWS Y	$\bar{Y}$	RELIGIOSITY X	$\bar{X}$	COMMUNITY REGRESSION LINES $y_{ic} = a_c + b_c x_{ic} + e_{ic}$
1	0	20	0	3	$y_{i1} = 10.9 + 3.0x_{i1} + e_{i1}$ $R^2 = 0.229$
	10		0		
	15		0		
	25		2		
	5		3		
	45		3		
	15		4		
	10		6		
	30		6		
	45		6		
2	10	30	1	4	$y_{i2} = 14.8 + 3.8x_{i2} + e_{i2}$ $R^2 = 0.424$
	20		1		
	30		1		
	20		2		
	15		4		
	45		4		
	25		6		
	30		7		
	50		7		
	55		7		
3	30	50	3	6	$y_{i3} = 18.1 + 5.3x_{i3} + e_{i3}$ $R^2 = 0.481$
	35		3		
	50		3		
	35		4		
	30		6		
	60		6		
	35		8		
	75		9		
	65		9		
	85		9		
4	35	70	4	7	$y_{i4} = 24.4 + 6.5x_{i4} + e_{i4}$ $R^2 = 0.445$
	35		4		
	80		4		
	70		6		
	40		7		
	95		7		
	75		8		
	85		10		
	90		10		
	95		10		

Values of the coefficients are

$$y_{is} = 8.27 + 6.84x_{is} + e_{is}, \qquad R^2 = 0.560 \tag{1.4}$$

For this larger sample there is a much stronger observed relationship between religiosity and opposition to reform. This is reflected in the value of R^2. The regression coefficient for religiosity is also larger for these data than for the first community.

Further examination shows that the relationships between opposition to abortion and religiosity are different in each community. The difference is shown on the right-hand side of Table 1.2. It is assumed that there are substantial cases in each community and that the observed differences are *not* due to sampling error. This suggests that community differences should be explicitly considered in the analysis of abortion-law attitudes. However, the analysis models considered so far treat only differences between individuals.

1.3

Explaining Variation in a Group Variable

The foregoing models focused on characteristics of individuals. However, Table 1.2 shows that communities differed substantially on levels of opposition. This is reflected in the community averages of Y, which measure a property of communities rather than individuals. This variable is denoted by $\bar{Y}$ to distinguish it from the individual variable Y. Given a concern for community differences, one might seek to explain the variation in this community variable in terms of other community variables. The level of religiosity in a community $\bar{X}$, reflected in the community averages of X, is a natural choice.

The relationship between $\bar{Y}$ and $\bar{X}$ is specified in the linear model

$$\bar{y}_c = a_g + b_g\bar{x}_c + e_c, \qquad c = 1, 2, \ldots, n_c \tag{1.5}$$

In Eq. (1.5) the subscript g indicates that the analysis involves groups (communities) rather than individuals. The residual term $\bar{e}_c$ refers to all other unspecified group variables. Examples could include socioeconomic levels, education levels, age or sex composition, and religious composition.

Computer procedures for this analysis are presented in Appendix A, Illustration 3. Estimates of the coefficients in Eq. (1.5) from the data in Table 1.2 are

$$\bar{y}_c = -17.5 + 12.0\bar{x}_c + e_c, \qquad R^2 = 0.976 \tag{1.6}$$

These results show that community-level religiosity is a strong predictor of community opposition to abortion liberalization. Provided that inferences about the observed relationship between group variables are not directed to individuals, the analysis is appropriate and can be useful in the broader study of abortion-law attitudes.[2]

There is, however, a problem with this approach. The objective of multilevel analysis treated in this chapter is to analyze variations in the individual variable Y, and not the group variable $\bar{Y}$. The analysis of the group-level variation should not be substituted for the analysis of the individual-level variation because it will generally produce different results. Compare the slope obtained from the group-level analysis to the slope obtained for each community separately and for the state sample. The differences demonstrate that the regression of group variables generally cannot be used to make inferences about individual characteristics. This problem is examined in Sections 2.2 and 10.1.

1.4

Analysis of Variance within and between Groups

Three approaches to the analysis of attitudes toward abortion reform are considered above. Two approaches involve the characteristics of individuals (for a general population and for specific subgroups in that population). The third involves only the characteristics of groups. Though fundamentally different in these respects, they have one important similarity. With each model the dependent and independent variables are observed at the *same* level (group or individual). These models are not sufficient by themselves to study the impact of individual *and* group differences on individual-level behavior.

An approach that explicitly considers individual and group effects is the conventional analysis of variance. This approach treats groups as categories of a nominal variable. The total variation in the dependent variable Y is partitioned into a within-group part and a between-group part. For consistency with subsequent discussion, the analysis of variance is represented in terms of the general linear regression model.[3]

The equation for the example is

$$y_{ic} = a_0 + a_2 z_{2ic} + a_3 z_{3ic} + a_4 z_{4ic} + e_{ic} \qquad (1.7)$$

In this equation y_{ic} refers to the abortion attitude score for the ith individual in the cth community. The z's are dummy variables indicating community residence. All

[2] The distinction between the incorrect use of group variables to make inferences about individuals and the legitimate use to make inferences about groups was stressed by Menzel (1950). Analyses of relationships between groups are often observed in the context of "cross-system comparative" studies. For a general discussion of this methodology, see Przeworski and Teune (1970).

[3] The application of the general linear model to traditional methods is discussed in Cohen (1968) and Fennessey (1968). For recent textbook treatments, see Namboodiri et al. (1975) and Kleinbaum and Kupper (1978).

individuals residing in community 2 will have a score of 1 on z_2 and a 0 on z_3 and z_4. To estimate coefficients when the model contains a regression intercept (a_0), a dummy variable for community 1 is excluded.

The value of a_0 is the value of the mean of Y in community 1, and the other coefficients equal the corresponding community means minus the intercept. The coefficients for the equation are

$$y_{ic} = 20.00 + 10.00z_{2ic} + 30.00z_{3ic} + 50.00z_{4ic} + e_{ic}, \qquad R^2 = 0.525 \qquad (1.8)$$

The average opposition score in community 1 is 20. The average in community 2 is 20 + 10 = 30. For communities 3 and 4 the averages are 50 and 70. Computer procedures for this analysis are presented in Appendix A, Illustration 4.

For the analysis of variance the between-community variation (sums of squared deviations of the within-group means of Y from the overall mean of Y) equals the regression sums of squares, or

$$\sum_c n_c(\bar{y}_c - \bar{y})^2 = \sum_c \sum_i (\hat{y}_{ic} - \bar{y})^2 = 14{,}750 \qquad (1.9)$$

where $\hat{y}_{ic}$ represents the predicted values of Y given Eq. (1.8). The within-community variation equals the residual sums of squares from the regression or

$$\sum_c \sum_i (y_{ic} - \bar{y}_c)^2 = \sum_c \sum_i (y_{ic} - \hat{y}_{ic})^2 = 13{,}350 \qquad (1.10)$$

The total variation is equal to the total sums of squares, or

$$\sum_c \sum_i (y_{ic} - \bar{y})^2 = 28{,}100 \qquad (1.11)$$

The ratio of the between-community variation to the total variation indicates that 14,750/28,100 = 52.5 percent of the variation in Y is accounted for by differences between communities. From the analysis of variance, it might be inferred that the within-community portion of the variance reflects the effects of individual-level variables, and the between-community portion reflects the effects of community-level variables.

This approach has several significant shortcomings. Individual-level effects are not identified with specific individual-level variables, and community effects are not identified with specific community variables. Without such a specification it is not possible to prevent mistaking one level of effects for the other. For example, the apparent group effect reflected in community differences in the means of Y may be due to an individual-level variable such as religiosity X. The analysis-of-variance approach also ignores the important possibility that individual-level effects vary from group to group, indicating an interaction between individual and group effects. The differences between the slopes relating abortion attitudes and religiosity in Table 1.2 demonstrate this possibility. Nevertheless, when theories relating group and individual variables are not well developed, this approach can be useful as a preliminary step to help identify potential sources of group effects.[4]

1.5

Analysis of Covariance Involving Group Differences and a Specified Individual-level Variable

Detecting spurious group-level effects requires an approach which controls for individual-level variables. The conventional analysis of covariance is one such approach.[5] Consider the specification of one individual-level variable which, in the example, is religiosity X. Instead of partitioning the variation of Y into a between- and a within-community part, as in the analysis of variance, the covariation between X and Y is divided into the two parts.

The covariation of X and Y is the sum of the products of the deviations of X and Y from their respective means, that is,

$$\sum_c \sum_i (x_{ic} - \bar{x})(y_{ic} - \bar{y}) = 2300 \tag{1.12}$$

The between-community covariation is obtained from the sum

$$\sum_c n_c(\bar{x}_c - \bar{x})(\bar{y}_c - \bar{y}) = 1200 \tag{1.13}$$

The within-community covariation is obtained from the sum

$$\sum_c \sum_i (x_{ic} - \bar{x}_c)(y_{ic} - \bar{y}_c) = 1100 \tag{1.14}$$

[4] The analysis-of-variance approach has been employed to identify group effects at successive layers of aggregation. This is referred to as "components analysis of variance." Kish shows that the variation on an individual-level variable can be divided into components associated with successive stages of subdivision, and that the ratio of each to the total variation can be taken as a measure of group effects originating at each level. He notes that the method involves measuring effects within units with no regard to their specific causes. He argues that the approach "can yield basic knowledge relevant to problems of social integration and organization; of differentiation and segregation, of homogeneity and diversity of types, classes and behaviors," and that "revealing the presence of homogeneity and its value can stimulate research into its meaning and origin." (Cf. Kish, 1962: 204–205.)

Stokes (1966: 61–85) extends the components analysis of variance for layered groups to include a component for time. In a study of election turnout and of party vote, total variation is divided into a district, a state, a national, and a time component. His conclusion is that local political subdivisions account for more individual-level variation in personal political behavior than the national unit.

Extensions of the basic model of contextual analysis to include more than one group type or level are considered in Chapter 7 of this book.

[5] For discussions of the analysis of covariance in the tradition of multilevel analysis, see Duncan, Cuzzort and Duncan (1961); Alker (1969); Schuessler (1969); Hauser (1970b); Werts and Linn (1971); Alwin (1976); Firebaugh (1978a).

The ratio of the within-community covariation to the total covariation indicates that $1100/2300 = 47.8$ percent of the total covariation of X and Y is accounted for within groups.

Partitioning the total covariation for X and Y loses sight of the objective of explaining variation in abortion attitudes in terms of community effects, while controlling for the individual effects of the individual-level variable, religiosity. Also, it does not distinguish between pure individual effects and effects due to an interaction between community- and individual-level effects.

By specifying a model containing terms for the group- and individual-level variables, it is possible to analyze variation in the dependent variable within the analysis-of-covariance framework. For the general linear model this involves introducing the specified individual variable X to the analysis-of-variance model in Eq. (1.7). The result is

$$y_{ic} = b_0 + b_2 z_{2ic} + b_3 z_{3ic} + b_4 z_{4ic} + b_5 x_{ic} + e_{ic} \tag{1.15}$$

Computer procedures are contained in Appendix A, Illustration 5. Estimates for this equation are

$$y_{ic} = 6.02 \ + 5.34 z_{2ic} + 16.02 z_{3ic} + 31.36 z_{4ic} + 4.66 x_{ic} + e_{ic}, \qquad R^2 = 0.707 \tag{1.16}$$

The value of R^2 in this model indicates that 70.7 percent of the total variation in Y is accounted for by community differences and the individual-level variable together. Compare this to 52.5 percent for the community variable alone and to 56.0 percent for the individual-level variable alone.

If an interpretation begins with the question of how much variation in Y is explained by the community variable after controlling for the individual variable X, the answer is $70.7 - 56.0 = 14.7$ percent. This might be taken to mean that a large portion of the apparent community effect observed in the simple analysis of variance was really due to the individual effect of X. However, one could as easily ask how much of the variation in Y is explained by the individual-level variable X after controlling for community differences. The answer is $70.7 - 52.5 = 18.2$ percent. This might suggest that an almost equally large portion of the apparent individual effect observed in the regression of Y on X alone was due to community differences.

A conclusion about which level is most important in explaining abortion attitudes therefore depends on which level is considered first. Interpreting the relative importance of explanatory variables is a problem since the variables are themselves related. This is the problem of multicollinearity which is encountered in virtually all multivariate analysis of nonexperimental data. Because of the explicit wish to separate group-level from individual-level effects, it is of special concern in multilevel analysis. Considerable attention is directed to potential remedies in Chapter 4.

The possibility of the individual-level effect being dependent on the community variable can also be studied within the analysis-of-covariance framework. An interaction between the community variable and the individual variable emerges as differences between the slopes relating Y and X across communities. These differences can be seen in Table 1.2.

By introducing interaction variables made from the products of the individual variable and each of the dummy variables in Eq. (1.6), this effect can be measured in a general linear model as follows:

$$\begin{aligned} y_{ic} = {} & b_0 + b_2 z_{2ic} + b_3 z_{3ic} + b_4 z_{4ic} + b_5 x_{ic} \\ & + b_6 (z_2 x)_{ic} + b_7 (z_3 x)_{ic} + b_8 (z_4 x)_{ic} + e_{ic} \end{aligned} \tag{1.17}$$

Estimates for this equation are

$$\begin{aligned} y_{ic} = {} & 10.89 + 3.95 z_{2ic} + 7.17 z_{3ic} + 13.48 z_{4ic} \\ & + 3.04 x_{ic} + 0.76 (z_2 x)_{ic} + 2.29 (z_3 x)_{ic} \\ & + 3.48 (z_4 x)_{ic} + e_{ic}, \qquad R^2 = 0.722 \end{aligned} \tag{1.18}$$

The value of R^2 in Eq. (1.18) indicates that 72.2 percent of the variation in abortion attitudes is explained by the model including dummy variables for the community effect, the individual variable involving religiosity, and the product variables for the interaction effect between the community- and individual-level variables.

Compare this to 70.7 percent for the model containing only the community- and individual-level variables [Eq. (1.16)]. The difference might be taken to mean that interaction effects account for only 1.5 percent of the variation in abortion attitudes. Such a conclusion can be misleading because of the problem of interpreting contributions of intercorrelated variables. For example, a model containing only terms for interaction accounts for 70.0 percent of the variation.

In terms of theory development, the analysis-of-covariance approach to the explanation of an individual-level variable is superior to the single-level approach described first, because it explicitly considers group effects. It is superior to the analysis-of-variance approach because it seeks to identify individual-level effects with specific individual-level variables. It is also more advanced in terms of theory construction because of the capability to detect the interaction of individual and group effects.

The major shortcoming of the analysis of covariance model is that it does not identify observed group-level effects with specific values of group variables. An analysis of covariance may find that community residence affects personal attitudes toward liberalizing abortion laws. However, a significant advance in substantive theory development would be to find that differences in the religious structure or composition of communities affect personal attitudes toward abortion reform. The analysis-of-covariance approach to the multilevel explanation of individual-level behaviors does not differentiate group-level effects.[6] It is most useful at early stages of theory development and in preliminary or exploratory analyses which seek to discover the existence of group-level effects and to find what groups are most relevant.

Technical problems arise when the number of groups in the analysis is large. If a study involved 80 or 100 units such as counties, specification and estimation of

[6] The analysis-of-covariance approach to group effects is compared to the contextual-analysis approach in Firebaugh (1978a).

the analysis-of-covariance model are complicated. However, this is increasingly less of a problem with advances in computer technology.

1.6 *Explaining Individual-level Variables with Specified Individual and Group Variables: Contextual Analysis*

The distinguishing characteristic of the approaches which are subsequently addressed is that group effects are tied to specific group variables. The general logic is to examine the effects of a specified individual-level variable, while controlling for the effects of a specified group variable and vice versa. This basic approach to the multilevel analysis of variation in individual behavior is referred to as "contextual analysis."[7] The label distinguishes it from the approaches previously considered and is consistent with related literature under that name.

1.6.1. The Blau Method for Categorized Variables

Early efforts to partial individual and group effects involved nominal or categorized individual-level variables and contingency tables.[8] To illustrate, consider the study of abortion attitudes. In Table 1.3 the individual-level scores on abortion attitudes and religiosity have each been dichotomized at their overall means. For personal religiosity, the categories are religious and nonreligious. For attitudes toward abortion reform, the categories are opposed and not opposed. They are tabulated to show the relationship between them.

Table 1.3 shows that 42.5 percent of all individuals opposed abortion reform. In the event of no relationship between religiosity and opposition to reform, the same proportion opposed could be expected in each category of religiosity. The observed difference between the two percentages ($65.0 - 20.0 = 45.0$) is a measure of the relationship between the two personal variables. For reference, the two (conditional) percentages are denoted by R and $\bar{R}$, referring to opposition given that a person is religious, and opposition given that a person is not religious. The difference ($R - \bar{R}$) is denoted by D.

Since this analysis involves all communities, it is similar to the approach described in Section 1.2. The differences are that the individual variables are

[7] The idea of contextual analysis is generally traced to Kendall and Lazarsfeld (1950); Blau (1960); and Davis et al. (1961). For more recent examples, see footnote 1, this chapter.
[8] Though the table method often is identified with Blau (1960), it was used earlier by Kendall and Lazarsfeld (1950).

Table 1.3
Categorized and Tabulated Data Showing the Relationship between Religiosity and Opposition to Abortion Reform for the State

OPPOSITION TO ABORTION REFORM	PERSONAL RELIGIOSITY		
	Not religious	Religious	Total
Opposed	4 (20.0)	13 (65.0)	17 (42.5)
Not opposed	16 (80.0)	7 (35.0)	23 (57.5)
Total	20 (100.0)	20 (100.0)	40 (100.0)

$R - \bar{R} = D$; $65.0 - 20.0 = 45.0$

categorized and the approach is not formally expressed as an equation. As in the first approach, there is no attempt to consider the effects of community residence on abortion attitudes.

One way to consider community differences is to construct a separate table for each community instead of one for all communities. Using the same overall means, the subscript c is used with the percentages R and $\bar{R}$ to identify community number. Observed differences between $\bar{R}_c$ across communities indicate undifferentiated community effects. Differences between the differences ($R_c - \bar{R}_c = D_c$) across communities indicate undifferentiated interactions between individual and community effects. Because there is no explicit identification of community effects with community-level variables, this is similar to the analysis-of-covariance approach described in Section 1.5.

Table 1.4 shows that there are considerable differences between communities. The percentages $\bar{R}_c$, referring to abortion attitude given that the individual is not religious, vary from 14.3 to 33.3 percent. The differences D_c vary from 19.0 to 52.4 percent.

One theory about specific sources of community effects is that "religious communities" influence personal attitudes toward abortion reform independent of personal religiosity. The percentages of religious people provide a basis for specifying the group variable. Communities are separated according to whether they are above or below the overall percentage of religious people. The percentages of religious people in the four communities are, respectively, 30.0, 40.0, 60.0, and 70.0. The overall percentage is 50.0. Observations for the first two communities are combined and treated as nonreligious communities. Observations for the second two communities are combined and treated as religious communities.

The data are retabulated in Table 1.5. The individual-level effect of religiosity is reflected in the relationship between religiosity and opposition for nonreligious

Table 1.4
Tables Showing Relationships between Religiosity and Abortion Reform within Communities

COMMUNITY	OPPOSITION TO ABORTION REFORM	PERSONAL RELIGIOSITY		
		Not Religious	Religious	Total
1	Opposed	1 (14.3)	1 (33.3)	2 (20.0)
	Not opposed	6 (85.7)	2 (66.7)	8 (80.0)
	Total	7 (100.0)	3 (100.0)	10 (100.0)
	$R_1 - \bar{R}_1 = D_1$; 33.3 − 14.3 = 19.0			
2	Opposed	1 (16.7)	2 (50.0)	3 (30.0)
	Not opposed	5 (83.3)	2 (50.0)	7 (70.0)
	Total	6 (100.0)	4 (100.0)	10 (100.0)
	$R_2 - \bar{R}_2 = D_2$; 50.0 − 16.7 = 33.3			
3	Opposed	1 (25.0)	4 (66.7)	5 (50.0)
	Not opposed	3 (75.0)	2 (33.3)	5 (50.0)
	Total	4 (100.0)	6 (100.0)	10 (100.0)
	$R_3 - \bar{R}_3 = D_3$; 66.7 − 25.0 = 41.7			
4	Opposed	1 (33.3)	6 (85.7)	7 (70.0)
	Not opposed	2 (66.7)	1 (14.3)	3 (30.0)
	Total	3 (100.0)	7 (100.0)	10 (100.0)
	$R_4 - \bar{R}_4 = D_4$; 85.7 − 33.3 = 52.4			

Table 1.5
Data for the Table Approach to Contextual Analysis with Categorized Variables

OPPOSITION TO ABORTION REFORM	COMMUNITY RELIGIOSITY: NONRELIGIOUS			COMMUNITY RELIGIOSITY: RELIGIOUS		
	Individual Religiosity			Individual Religiosity		
	Not Religious	Religious	Total	Not Religious	Religious	Total
Opposed	2	3	5	2	10	12
	(15.4)	(42.9)	(25.0)	(28.6)	(76.9)	(60.0)
Not opposed	11	4	15	5	3	8
	(84.6)	(57.1)	(75.0)	(71.4)	(23.0)	(40.0)
Total	13	7	20	7	13	20
	(100.0)	(100.0)	(100.0)	(100.0)	(100.0)	(100.0)

communities. The subscripts r and nr are used with R and $\bar{R}$ to differentiate the set of religious communities and the set of nonreligious communities. In the table for nonreligious communities, the difference ($R_{nr} - \bar{R}_{nr} = 42.9 - 15.4 = 27.5$) is a measure of the individual effect of religiosity.

The community effect identified with the specified group variable is reflected in the difference of percentages shown in the opposed/nonreligious cells of the table between religious and nonreligious communities. The difference ($\bar{R}_r - \bar{R}_{nr} = 28.6 - 15.4 = 13.2$) is a measure of the community effect associated with community religiosity. A comparison of this difference to the difference involving the opposed/religious cells ($R_r - R_{nr} = 76.9 - 42.9 = 34.0$) indicates the existence of an interaction between individual and community effects. The community effect is stronger for religious than for nonreligious individuals. The interaction effect is also recognized in the difference of relationships involving religiosity and reform opposition between religious and nonreligious communities. The difference ($42.9 - 15.4 = 27.5$) measures the relationship in the nonreligious set of communities. The difference ($76.9 - 28.6 = 48.3$) measures the relationship in the religious communities. Therefore the individual effect of religiosity on abortion attitudes is stronger in religious communities than in nonreligious communities.

An early precedent for the multilevel approach to the explanation of an individual-level variable where the group variable is specified was reported in the study of public welfare agencies by Blau (1960). The object of the analysis was to explain why some caseworkers looked upon their jobs as eligibility determination and others as social service provision. The individual-level explanatory variable involved caseworker attitudes (positive or negative) toward client recipients. The group-level variable was defined as the *prevailing* attitude of workers toward clients in agencies (predominantly positive or negative). According to Blau, "The structural effects of a social value can be isolated by showing that the association between its prevalence in a community or group and certain patterns of conduct is independent of whether individuals hold the value or not" (Blau, 1960:178).

Table 1.6
Blau's Data Showing the "Structural Effects of Social Values"

ATTITUDES TOWARD WORK	PREVAILING ATTITUDES TOWARD CLIENTS IN AGENCIES: POSITIVE		NEGATIVE	
	Individual Attitudes toward Clients		Individual Attitudes toward Clients	
	Positive	Negative	Positive	Negative
Eligibility determination	30	56	56	55
Intermediate	10	0	0	18
Providing services	60	[44]	44	[27]
	100%	100%	100%	100%

Source: Adapted from Blau, 1960, p. 181.

Data for this analysis are tabulated in Table 1.6. The first set of agencies consists of those predominantly positive in attitudes toward clients. The second consists of those predominantly negative. The agency effect associated with the group variable is observed in the pair of bracketed percentages. The difference (44 – 27 = 17) is a measure of the agency effect of prevailing attitudes toward clients on personal job perceptions. The individual effect is observed in the pair of percentages enclosed by the dashed lines. The difference (60 – 44 = 16) is a measure of the individual effect of personal attitudes toward clients on personal job perceptions. The difference for the same categories in the predominately negative agencies is 44 – 27 = 17. The small difference between the two (namely, 16 and 17) indicates that there is no substantial interaction between individual and group effects.

The primary virtue of this approach compared to the analysis of covariance is its ability to differentiate group effects and to identify them with specific group-level variables. The approach has a special appeal because of its simplicity and intuitive ring. But it also has several significant limitations. A major shortcoming results from categorizing continuous individual-level variables (Tannenbaum and Bachman, 1964). The selection of cutpoints is often arbitrary and can produce misleading evidence about the existence of individual and group effects. When the group variable is based on group proportions or percentages, it too is categorized by the Blau method. This operation can also lead to deceiving conclusions about the presence of group- and individual-level effects.

1.6.2. Contextual Analysis with Categorical Individual-level Variables and a Continuous Group-level Variable: The Davis Method

A significant advance over the Blau method is contained in the technique described by Davis et al. (1961).[9] Instead of collapsing observations on a continuous group-

[9] A more recent example of its use is in Bowers (1968).

level variable, the value for each group is preserved. In the study of abortion reform, Table 1.4 shows that the percentage of individuals being religious are 30.0, 40.0, 60.0, and 70.0. Instead of having religious and nonreligious communities, degrees of community religiosity are maintained.

The method directs attention to the relationships between the group variable and the within-community percentages of individuals opposed to reform for each category of personal religiosity. Let R be the percentage opposed to abortion reform among religious individuals. In community 1, one out of three religious individuals opposed abortion reform. The percentage R_1 therefore equals 33.3. The corresponding conditional percentages for the other communities are $R_2 = 50.0$, $R_3 = 66.7$, and $R_4 = 85.7$. Let $\bar{R}$ be the percentage opposed to abortion reform among nonreligious individuals. In community 1, one out of seven nonreligious individuals opposed abortion reform, and $\bar{R}_1$ therefore equals 14.3. The other percentages, $\bar{R}_2$, $\bar{R}_3$, and $\bar{R}_4$, are given in Table 1.4.

The relationships between R and P, and between $\bar{R}$ and P, are visualized in the plot shown in Figure 1.1. The interpretation of multilevel effects is made from this graph. The difference between R and $\bar{R}$ at low values of P indicates the individual effect of religiosity on abortion attitudes. The increase in R and $\bar{R}$ with P indicates

Figure 1.1
Graph of Multilevel Effects with Categorized Individual-level Variables and a Continuous Group-level Variable: The Davis Method

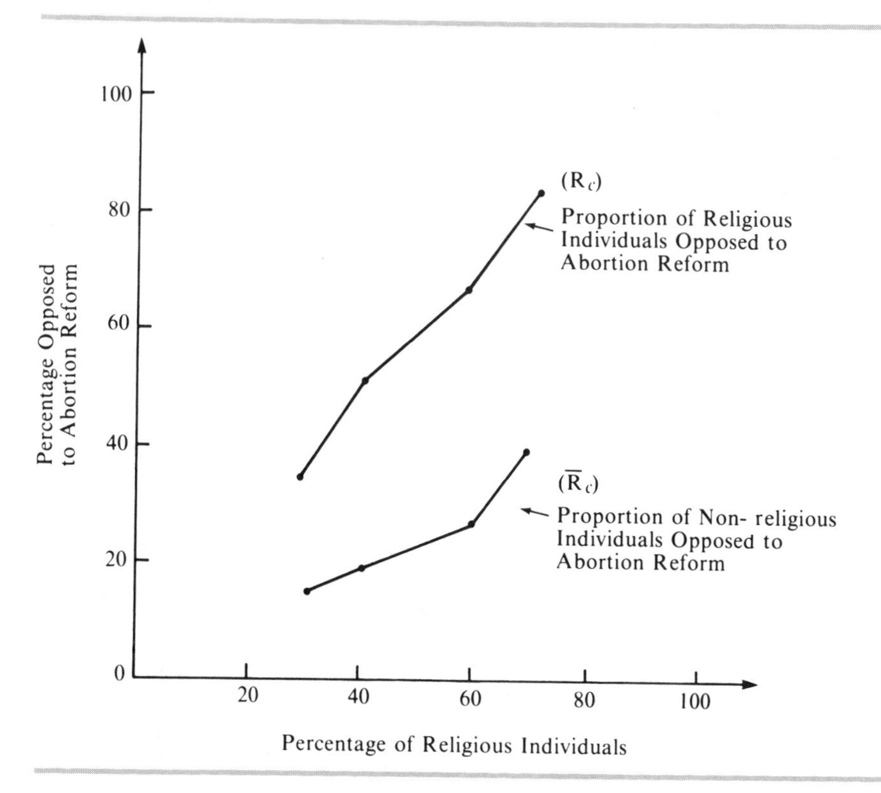

the group effect identified with community religiosity. The difference between R and $\bar{R}$ at low and high values of P indicates an interaction between individual and group effects.

The major advantage of the Davis method is that it uses all the information in a continuous group-level variable. The visual and intuitive representation of multilevel effects is effective with a small set of variables where theory development seeks to identify complex combinations of individual, group, and interaction effects. Davis identifies several such possibilities. For example, consider the example drawn from Davis in Figure 1.2.

Figure 1.2
Relationship between Attendance and Active Participation in Great Books Clubs.

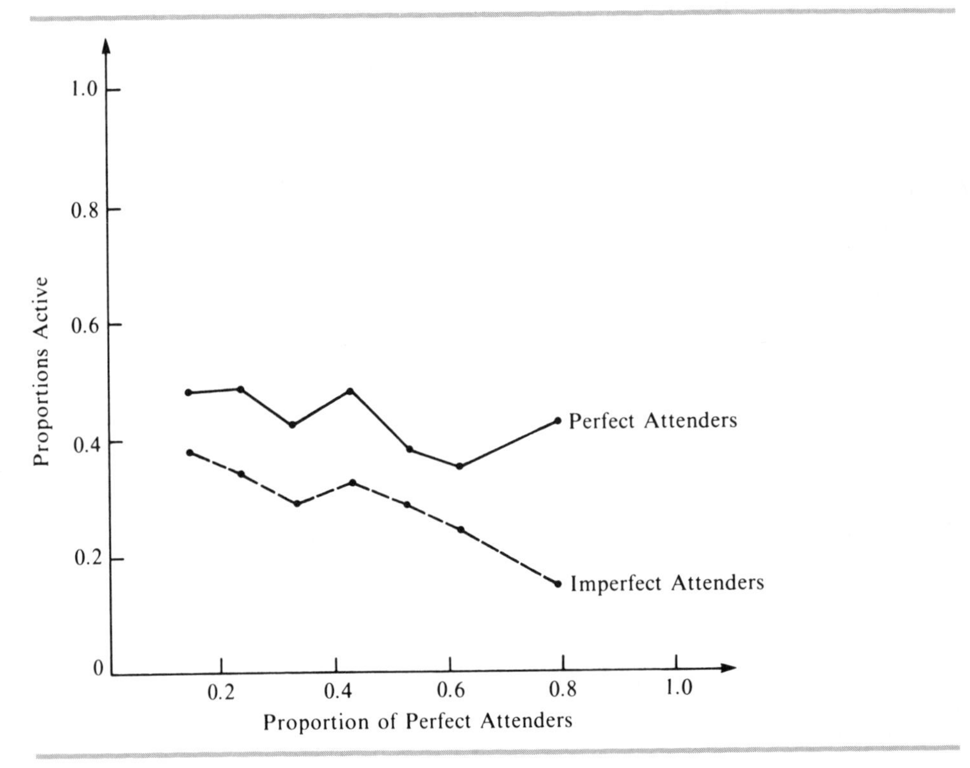

Source: Davis et al., 1961, p. 225.

This analysis examined the relationship between perfect and imperfect attenders of great books clubs and active participation in meetings. The group variable is level of active participation. The graph indicates that the relationship between attendance and active participation is positive. However, as the level of activism increases, active participation for perfect *and* imperfect attenders decreases. This illustrates that individual and group effects can operate in opposite directions in what is referred to as a "boomerang effect" (cf. Davis et al., 1961 : 224–225).

The approach also has significant limitations. The problem of categorizing continuous individual-level variables remains, as does the consequent potential for confusing individual and group effects. Like the Blau method, it does not formally represent the analysis mathematically so that the variation of individual-level behavior can be studied in terms of individual and group effects. The analysis of covariance approach is superior in this respect, but inferior in that it cannot differentiate group effects in terms of specific group-level variables. The Davis approach is also constrained to the treatment of only one individual-level and one group-level variable. This is a serious limitation in substantive areas where substantive theory is advanced.

1.7
The Basic Model of Contextual Analysis with Continuous Individual and Group Variables

The specific limitations of previous approaches to the multilevel analysis of individual behavior identified above are remedied with the specification of a multiple-regression model containing continuous individual, group, and interaction variables.[10] This model is

$$y_{ic} = b_0 + b_1 x_{ic} + b_2 \bar{x}_c + b_3 x_{ic} \bar{x}_c + e_{ic} \tag{1.19}$$

As in the analysis of covariance, y_{ic} and x_{ic} represent continuous individual-level scores on opposition to reform and religiosity for the ith individual in the cth community. The variable $\bar{x}_c$ represents community-level religiosity. The possibility of an interaction between personal religiosity and community religiosity is addressed with the product variable $x_{ic}\bar{x}_c$. The partial coefficients b_1, b_2, and b_3 measure the individual, group, and interaction effects associated with these variables.

Computer procedures are contained in Appendix A, Illustration 6. Estimates for this equation, given the data in Table 1.2, are

$$y_{ic} = 4.08 - 0.14 x_{ic} + 2.54 \bar{x}_c + 0.96 x_{ic} \bar{x}_c + e_{ic}, \qquad R^2 = 0.717 \tag{1.20}$$

The value of R^2 indicates that the model containing specified explanatory variables accounts for 71.7 percent of the variation in opposition to abortion reform. This compares to 72.2 percent for the analysis-of-covariance model containing unspecified group and interaction variables. The difference is that the nominal variable treatment of community effects in the analysis-of-covariance model detects a composite of all

[10] This formulation appears in Boyd (1971); Hanushek et al. (1974); Przeworski (1974); Alwin (1976); and Firebaugh (1978a).

community effects without differentiating them in terms of specific community-level properties.[11] In contrast, the context model detects the specific community effect associated with community religiosity. This point is elucidated in Chapter 4.

From the numerical values in Eq. (1.20) one might infer that substantial community and interaction effects are associated with community religiosity and that essentially no pure individual effect is identified with personal religiosity. Note that the sign of the partial coefficient for the individual variable is negative. After controlling for community and interaction effects, opposition to abortion liberalization decreases with personal religiosity.

As in any model simultaneously treating individual- and group-level variables, interpreting the relative importance of explanatory variables in this model is problematic. Once again the variables are intercorrelated. Conclusions about the unique contributions of each to the explanation of individual behavior therefore depend on the order in which they are considered. When the individual variable is considered first, the regression coefficient is positive rather than negative, and the variable alone accounts for 56.0 percent of the variation in opposition to abortion reform. This leaves 14.7 percent accounted for by the community and interaction variables together. However, when the community and interaction variables are considered first, virtually nothing is left for the individual-level variable to explain. This problem is pursued in Chapters 4 and 13.

Because of the importance of the contextual model, it is thoroughly examined in Chapter 3. Besides the problem of multicollinearity common to all multilevel approaches to the explanation of individual behavior, the simple context model has other limitations. These limitations are explicitly addressed in Chapter 13 and are encountered throughout.

1.8
Summary

This chapter reviews and examines the relative merits of several approaches to the study of individual behavior. The contextual model is considered superior to models which ignore group effects in the explanation of individual-level variables. It is an advance over the analysis-of-variance approach because it identifies individual and group effects with specific individual and group variables. It is an advance over the analysis-of-covariance model because that approach does not identify group effects with specific group variables. The formalized contextual model treated last is considered superior to earlier contextual analysis approaches (for example, the Blau and Davis methods) because it does not require categorized variables. Being mathematically expressed, the contextual model generalizes more easily to complex substantive theories involving nonlinear relationships between variables, more than one individual- or group-level variable, and more than one kind of group. These extensions are pursued in Chapters 6 and 7.

[11] This point is stressed in Firebaugh (1978a).

Though the contextual model is considered more advanced, the analysis-of-variance and -covariance approaches can be useful when substantive theory about contextual effects is not well developed. Also, the simple contextual model has limitations that require attention in applying the method. Attention is directed, for example, to the problem of correlated explanatory variables.

Another important contribution of the contextual model goes beyond the objective of explaining individual behavior as an end in itself. This involves its potential to further understand invariant properties of a single unit as observed in regression coefficients relating individual- or group-level variables in a single unit such as a state or nation. This whole subject is introduced as one of the major objectives of multilevel analysis and is pursued in depth in Chapters 8 and 9.

Introduction to the Second Objective of Multilevel Analysis: Explaining Aggregate Relationships

This chapter refers to the data and analyses described in Chapter 1 with a different multilevel research problem in mind. Instead of seeking to explain individual behavior, the objective is to account for the relationship between community-level religiosity and community-level opposition to abortion reform for the state. This example illustrates a simple but methodologically explicit step toward integrating micro and macro sociology.[1] Also introduced is the related problem of using group-level data to study individual behavior. This is the problem of "cross-level inference."[2] The purpose of this chapter is to *preview* the second objective of multilevel analysis. The subject is extensively discussed in Part III of the book.

[1] The challenge to synthesize micro and macro perspectives on social research is discussed by Blalock (1967); Coleman (1970); and Boudon (1969). Allardt (1968) describes contextual analysis as a synthesis of American micro and European macro traditions in research. Related discussions begin with criticisms of strict micro-level research, for example, Barton (1968) and Price (1968). A central philosophy-of-science concept behind multilevel analysis is "analytical reduction." The essence is that understanding macro-level phenomena, such as national housing patterns, requires the analysis of smaller units, such as groups and individuals. Lipset et al. (1956: 91–132); Nagel (1961: 345–358, 541–544); Stokes (1966); Eulau (1969: chap. 1).

[2] A sampling of literature involving cross-level inference is Duncan and Davis (1953); Goodman (1953, 1959); Hammond (1973); Irwin and Meeter (1969); Iversen (1969, 1973); Kousser (1973); Meckstroth (1970, 1974); Robinson (1950); Shively (1969, 1974); Slatin (1969); Stokes (1969). A recent addition to this literature is Firebaugh (1978b). Discussion relevant to cross-level inference is also found with treatments of "the ecological fallacies" and "the problems of aggregation." Examples include Alker (1969); Green (1964); Grunfeld and Griliches (1960); Hannan (1971a,b); Hanushek et al. (1974); Orcutt et al. (1968).

2.1

Explaining a Relationship between Group Averages with a Contextual Model of Individual-level Behavior

In Table 1.2 the regression coefficients relating abortion opposition Y and religiosity X within communities are 3.0, 3.8, 5.3, and 6.5. The coefficient b_s [from Eq. (1.4)] for the state is 6.8, and the coefficient [from Eq. (1.6)] relating the community averages $\bar{y}_c$ and $\bar{x}_c$ is 12.0. These are puzzling findings. They apply to the same substantive problem, but they are numerically very different. Not only do the within-group results vary substantially, the *average* of the community coefficients (4.6) does not equal the coefficient for the state. Unequal community size is no explanation because the communities are equal in size and no community has a coefficient larger than the state coefficient. Even more puzzling is the observation that the regression coefficient for the averaged variables is considerably larger than the within-community coefficients *and* the state coefficient.

The general problem is to explain why such differences occur. This question involves the related issues of cross-level inference and ecological fallacies considered later in this chapter and again in Chapter 10. In this chapter the question is addressed to account for a group-level relationship observed between two variables in terms of individual behavior. Next a more complex example is introduced to suggest the scope of this multilevel objective. Chapter 8 shows how a population relationship, rather than a group-level relationship, is explained through contextual analysis.

In this context the meaning of "explain" is different than in Chapter 1. When the objective is to "explain" individual behavior, the meaning involves the analysis of variation in a dependent individual-level variable. When the objective is to "explain" an observed relationship in a single unit, the meaning involves aggregating relationships between subparts of the single unit to reproduce the observed relationships between subparts of the single unit to reproduce the observed relationship.[3] Though the two kinds of explanation are distinctively different, the first is prerequisite to the second.

It is observed in Chapter 1 that the relationship between opposition Y and religiosity X for the general population (the state s) can be studied using the regression model

$$y_{ic} = a_s + b_s x_{ic} + e_{ic} \tag{2.1}$$

The subscripts i and c identify each individual person and the person's community.

Instead of estimating the coefficients in Eq. (2.1) directly, observations on X, Y, and the residual variable can be summed for each community to obtain community averages. For example, community 1 has $\bar{y}_1 = 20$.

[3] Excellent discussions of philosophy-of-science principles germane to the multilevel perspective are contained in Eulau (1969: chap. 1); Lipset et al. (1956: 91–132); Stokes (1966).

Eq. (2.1) then can be written

$$\bar{y}_c = a_s + b_s \bar{x}_c + \bar{e}_c \tag{2.2}$$

Eq. (2.2) has the same form as Eq. (1.5), which was used to explain variation in the group variable $\bar{Y}$. The coefficients in Eq. (1.5) are denoted by a_g and b_g to indicate that the model involves the analysis of group variables. At first one might intuitively expect the numerical values of a_s and b_s to be the same as the regression estimates obtained from Eq. (1.5), particularly because the two equations seem to have the same form. Instead, the estimates for a_s and b_s, obtained by regressing Y on X for all 40 observations in Table 1.2, are 8.3 and 6.8. The estimates for a_g and b_g, obtained by regressing the means of Y on the means of X for the four communities in Table 1.2, are -17.5 and 12.0. Why are they different? The answer accounts for the shape of the observed group-level relationship and, in this sense, explains it.

The two sets of regression coefficients differ because the model of individual behavior aggregated to the community level is incorrectly specified. The group effect associated with community religiosity $\bar{X}$ on abortion opposition has been ignored. By including the group effect, the model can be expressed by the equation

$$y_{ic} = b_0 + b_1 x_{ic} + b_2 \bar{x}_c + e_{ic} \tag{2.3}$$

Estimates for Eq. (2.3) are

$$y_{ic} = -17.5 + 4.7 x_{ic} + 7.3 \bar{x}_c + e_{ic} \tag{2.4}$$

When observations are summed within communities and divided by community size n_c, we get

$$\bar{y}_c = b_0 + b_1 \bar{x}_c + b_2 \bar{x}_c + \bar{e}_c \tag{2.5}$$

This equation reduces to

$$\bar{y}_c = b_0 + (b_1 + b_2)\bar{x}_c + \bar{e}_c \tag{2.6}$$

Eq. (2.6) involves only the group averages of X and Y. This is also the case for Eqs. (1.5) and (2.2). Substituting the estimates obtained in Eq. (2.4) into Eq. (2.6) gives

$$\begin{aligned} \bar{y}_c &= -17.5 + (4.7 + 7.3)\bar{x}_c + \bar{e}_c \\ &= -17.5 + 12.0\bar{x}_c + \bar{e}_c \end{aligned} \tag{2.7}$$

Thus the regression coefficients in Eq. (2.7), based on the contextual model of individual behavior, are identical to the coefficients obtained directly from the regression of the community averages $\bar{y}_c$ on the averages $\bar{x}_c$ in Eq. (1.6). We discover that the coefficient b_g relating the group variables is made up of two parts. One of these reflects the individual-level effect of X. The other reflects the group effect of $\bar{X}$.

When the two effects operate in the same direction, the slope relating the group means is larger than the slope relating X and Y within groups and in the general population. It is larger than the individual effect in the amount of the group effect. In the example, this amount ($12.0 - 4.7 = 7.3$) is over half the numerical value of the observed group-level coefficient.

Finding individual and group effects with opposite signs is also conceivable. If they are relatively close in magnitude, they will tend to cancel out. This explains some situations in which one expects a significant relationship between two group variables but none is observed.

The meaning of "explanation" in this kind of multilevel analysis can be summarized as follows. When regression coefficients of the group-level equation are reproduced by aggregating a model of individual behavior, the aggregate relationship has been explained. This occurs when the underlying model of behavior, involving individual and contextual variables, has been correctly specified.

The above example is a simple case of the broader and longer range objective of social science to understand complex societal-level phenomena in terms of the motives and behaviors of individuals. In the example the single unit was a state. The aggregate property to be explained was the relationship between community-level opposition to abortion reform and community-level religiosity. Yet explanations of aggregate phenomena involving whole nations and other complex systems are needed. Phenomena needing explanation are as sweeping as social movements, economic curves, political structures, political outcomes, and patterns of segregation and discrimination. The methodology of achieving that understanding is complex and not yet well defined or understood. However, the strategy observed in the simpler case described above suggests the basic logic of this methodology.

2.1.1. A More Complex Example

To illustrate this logic with a real and somewhat more complex example, consider a national political phenomenon known as the "puzzle of uniform election swings in Britain" (cf. Butler and Stokes, 1969). The aggregate phenomenon is that the percentage change in the Conservative (or Labour) vote from one election to the next in Britain is very nearly the same in each of the 631 local British constituencies. The percentage change is known as the "swing." The puzzle is that the swings across constituencies are almost constant. How is this explained in terms of individual behavior?

An old and popular explanation is that voters depart from strict party allegiance when they are sufficiently influenced by events. Individuals in homogeneous Britain are influenced by the *same* (national) events. Therefore, departures from party allegiance are constant across constituencies and so, too, are the swings. But is this popular explanation valid? This question can be pursued as follows.

The phenomenon of election swings can be represented by the aggregate-level equation

$$\bar{y}_c = a_g + b_g\bar{x}_c + e_c \tag{2.8}$$

In Eq. (2.8), $\bar{x}_c$ represents the proportions of votes in a national election for the Conservative party in constituency c. The proportion of votes for the Conservative party in the following national election for the same constituency is denoted by $\bar{y}_c$. The subscript g, used with the coefficients, indicates that they refer to a group-level equation. If there were no swings from one election to the next, the slope in Eq. (2.8) would be 1 and the intercept would be 0. A constant swing across all constituencies gives a nonzero intercept and a slope equal to 1. If the swing is not constant, the slope may differ from 1. The uniformity of swings will be observed in the fit of the regression line to the constituency data. Our objective is to account for the relationship observed in Eq. (2.8). This will have been accomplished when that relationship is reproduced by aggregating the appropriate model of individual voting behavior.

Implicit in the popular theory described above is a model of individual behavior which does not take into account group effects due to differences in the partisan composition in constituencies. The equation for this model is of the same form as the noncontext model specified in Section 1.2:

$$y_{ic} = a_p + b_p x_{ic} + e_{ic} \tag{2.9}$$

In the equation, X and Y are dummy variables specifying the vote for the Conservative or Labour party for the two successive elections. The subscript p makes clear that they apply to a regression involving the (national) population. The subscript ic identifies the ith person in the cth constituency. When the individual variables in Eq. (2.9) are aggregated to the level of constituencies by summing them within constituencies and dividing by the corresponding n's, we get

$$\bar{y}_c = a_p + b_p \bar{x}_c + \bar{e}_c \tag{2.10}$$

This equation now has the same form as Eq. (2.8) and represents the relationship to be explained. If the estimates of the coefficients in Eq. (2.10), which were obtained from the individual-level model in Eq. (2.9), are the same as the estimates obtained from Eq. (2.8), then the aggregate phenomenon is consistent with the individual-level theory said to lie behind it.

This example is pursued at length in Chapter 12. There it is demonstrated that the popular explanation of the phenomenon of uniform election swings is not valid. As in the case described earlier, the correctly specified model of individual behavior explicitly considers the group effect associated with the group variable. In this case the partisan composition of constituencies is involved.

The example illustrates the two kinds of explanations involved when the objective is to understand aggregate phenomena in terms of individual behavior. The first kind of explanation involves relationships between individual-level variables. The dependent variable is said to be explained to the extent that its variation is accounted for by the independent variables. The second kind of explanation involves observed relationships between group-level variables. These relationships are explained to the extent that they are reproduced by aggregating the individual-level relationships to the group level.

2.2

Reconstructing Missing Individual-level Information to Study Individual-level Behavior: Cross-level Inference

In each of the two previous analyses, the focus is on a relationship between two group-level variables. The objective is to reproduce the coefficients of that relationship by aggregating a correctly specified model of individual-level behavior in a regression analysis. When this is the objective, there is usually some direct theoretical interest in the observed group-level relationship. The relationship might, for example, be larger or smaller in magnitude than theory or prior research would indicate. As in the example of national election outcomes, we might seek to understand some momentous aggregate-level phenomenon. There are occasions, however, when researchers become interested in group-level data even when their objective is to explain individual-level behavior, and when there is no direct theoretical interest in group-level relationships. This happens when the *only* data available involve observations at the group level such as means and proportions.

Following the example used in Chapter 1, imagine that the objective is to study individual-level opposition to abortion reform in terms of personal religiosity. But instead of (self-reported) survey data on these variables, we have aggregate data consisting of the percentages of individuals who voted against local abortion-law referenda. From a separate source we have the percentages of individuals who attend church. Assume that the reform data are obtained from election results for districts whose borders correspond to the four communities in the previous example. The religiosity data are obtained from census data or from another study. Also, imagine that both variables are conceptually the same as used in the previous example.

The group-level observations for the four communities are presented in Table 2.1. To develop the point of cross-level inference, they are shown in Table 2.2 as margins of contingency tables for each community. These tables are identical to

Table 2.1
Group-level Observations for Four Communities

COMMUNITY NUMBER	FREQUENCY AND PERCENTAGE OF RELIGIOUS PEOPLE	FREQUENCY AND PERCENTAGE OF PEOPLE OPPOSED TO REFORM
1	3 (30)	2 (20)
2	4 (40)	3 (30)
3	6 (60)	5 (50)
4	7 (70)	7 (70)
Total	20 (50)	17 (42.5)

Table 2.2
Observations for Margins with Cell Observations Missing

COMMUNITY

1

	Not religious	Religious	
Opposed to reform			2 (20.0)
Not opposed to reform			8 (80.0)
	7 (100.0)	3 (100.0)	10 (100.0)

2

	Not religious	Religious	
Opposed to reform			3 (30.0)
Not opposed to reform			7 (70.0)
	6 (100.0)	4 (100.0)	10 (100.0)

3

	Not religious	Religious	
Opposed to reform			5 (50.0)
Not opposed to reform			5 (50.0)
	4 (100.0)	6 (100.0)	10 (100.0)

4

	Not religious	Religious	
Opposed to reform			7 (70.0)
Not opposed to reform			3 (30.0)
	3 (100.0)	7 (100.0)	10 (100.0)

those presented in Table 1.4 for the example with complete survey data, except for one crucial difference. The *joint distributions* of the two variables which should appear in the cells of the tables are missing. The cell information in conjunction with the marginal information allows inferences to be made about the individual-level relationship between the two variables. Specifically, a relationship is indicated when the percentages opposed to reform *within* the categories of religiosity depart significantly from the percentage opposed to reform observed in the margin.

The general problem of cross-level inference can be stated in terms of this observation. How can the joint distribution information in the cells of Table 2.2 be *recovered* to allow inferences about individual behavior (cf. Iversen, 1969, 1973).[2]

Early approaches sought to recover missing individual-level information exclusively from group-level information. One approach simply substituted the group data for the individual data.[4] To illustrate, the relationship between community-level opposition to reform and community religiosity can be studied with the regression model

$$\bar{y}_c = a_c + b_c\bar{x}_c + e_c \tag{2.11}$$

where $\bar{y}_c$ and $\bar{x}_c$ represent the community percentages of people opposed to reform and the percentages of church attenders. Observations on these variables are contained in Table 2.1.

Table 2.3
Differences in Conditional Percentages and Correlations for the Four Communities and the State

COMMUNITY	DIFFERENCE IN CONDITIONAL PERCENTAGES	DIFFERENCE IN CORRELATION COEFFICIENTS
1	19	0.22
2	33	0.36
3	42	0.41
4	52	0.52
Total (state)	45	0.46

Estimates for Eq. (2.11) are

$$\bar{y}_c = -17.5 + 12.0\bar{x}_c + e_c, \qquad R^2 = 0.976 \tag{2.12}$$

The coefficients from Eq. (2.12) for group data are applied to individual behavior. From the value of R^2 it is inferred that individual-level religiosity explains 97.6 percent of the variance in opposition to abortion-law reform.

To illustrate the extent to which this procedure can mislead, assume that the missing joint distributions of the individual-level variables in Table 2.2 are the same

[4] Studies which used group data to make inferences about individuals were said to have committed the "ecological fallacy." (Cf. Robinson, 1950; Riley, 1964; Alker, 1969; Stokes, 1969.)

Table 2.4
Minimum and Maximum Possible Differences in Percentages and Correlation Coefficients for Four Communities and the State, Given Only Margin Data

COMMUNITY	LOWER LIMIT				UPPER LIMIT			
1		$\bar{R}$	R			$\bar{R}$	R	
	Op	0	2	2	Op	2	0	2
		(00.0)	(66.7)			(28.6)	(00.0)	
	$\overline{Op}$	7	1	8	$\overline{Op}$	5	3	8
		(100.)	(33.3)			(71.4)	(100.)	
		7	3	10		7	3	10
	$D_1 = 66.7,\quad r_1 = 0.76$				$D_1 = -28.6,\quad r_1 = -0.33$			
2		$\bar{R}$	R			$\bar{R}$	R	
	Op	0	3	3	Op	3	0	3
		(00.0)	(75.0)			(50.0)	(00.0)	
	$\overline{Op}$	6	1	3	$\overline{Op}$	3	4	7
		(100.)	(100.)			(50.0)	(100.)	
		6	4	10		6	4	10
	$D_2 = 75.0,\quad r_2 = 0.80$				$D_2 = -50.0,\quad r_2 = -0.54$			
3		$\bar{R}$	R			$\bar{R}$	R	
	Op	0	5	5	Op	4	1	5
		(00.0)	(83.3)			(100.)	(16.7)	
	$\overline{Op}$	4	1	5	$\overline{Op}$	0	5	5
		(100.)	(16.7)			(00.0)	(83.3)	
		4	6	10		4	6	10
	$D_3 = 83.3,\quad r_3 = 0.82$				$D_3 = -83.3,\quad r_3 = -0.82$			
4		$\bar{R}$	R			$\bar{R}$	R	
	Op	0	7	7	Op	3	4	7
		(00.0)	(100.)			(100.)	(57.1)	
	$\overline{Op}$	3	0	3	$\overline{Op}$	0	3	3
		(100.)	(00.0)			(00.0)	(42.9)	
		3	7	10		3	7	10
	$D_4 = 100.0,\quad r_4 = 1.0$				$D_4 = -42.9,\quad r_4 = -0.43$			
Sum table (state)		$\bar{R}$	R			$\bar{R}$	R	
	Op	0	17	17	Op	12	5	17
		(00.0)	(85.0)			(60.0)	(25.0)	
	$\overline{Op}$	20	3	23	$\overline{Op}$	8	15	23
		(100.)	(15.0)			(40.0)	(75.0)	
		20	20	40		20	20	40
	$D_s = 85.0,\quad r_s = 0.86$				$D_s = -35.0,\quad r_s = -0.35$			

as in Table 1.4. The dichotomous data equivalents to the slopes and correlation coefficients for each of the communities are shown in Table 2.3. The group-level correlation Eq. (2.12) greatly exaggerates the relationship between individual-level opposition to reform and religiosity.

This problem was first observed in Section 1.2 in the discussion of explaining variance in group variables. The reason for the problem is contained in the discussion following Eq. (2.7). The coefficient relating group-level variables is the consequence of individual *and* group effects on the individual-level dependent variable. When the two effects operate in the same direction, the result of the group-level analysis will overestimate the individual effect.

Another approach to recovering missing joint distribution data with group data only is to compute the upper and lower *limits* of the individual-level relationships.[5] To illustrate this approach, consider the marginal frequencies for community 1 in Table 2.3. Because the cell frequencies must add up to the corresponding marginal frequencies, the two extreme possibilities can be identified. These two possibilities for community 1 are shown in Table 2.4. The frequency in the religious/opposed cell cannot be greater than the total number opposed or the total number religious. This frequency is 2. The smallest frequency possible for that cell is 0.0. The two limiting cases for each community are shown in Table 2.4 along with the minimum and maximum differences in percentages D_c and the minimum and maximum correlation coefficients r_c. The range of possible values of r_c for community 1 is -0.33 to $+0.76$. The range of possible values for community 3 (-0.83 to $+0.83$) is even larger. The example illustrates the shortcomings of this approach. Limits obtained from the margins alone may be so large as to be of little use. Other attempts to recover missing joint-distribution data are examined in Chapter 10 in the context of the more general statement of the problem, the study of individual behavior with incomplete data.

2.3 *Summary*

This chapter discusses a multilevel research enterprise that is different from (though dependent on) contextual analysis. The objective is to explain group-level relationships in terms of individual-level models of behavior containing group variables. This was described as a simple but methodologically explicit case of blending micro and macro sociology. The related problem of cross-level inference is also introduced. In cross-level inference there also is an interest in group-level relationships, but only because that is the only information at hand to study individual-level relationships.

The purpose of this chapter is to introduce the second objective of multilevel analysis: explaining aggregate relationships. This subject is more fully pursued in Part III of the book. Cross-level inference is discussed more fully in Part IV.

[5] This procedure was formalized by Duncan and Davis (1953).

Explaining Individual-level Relationships

Two major objectives of multilevel analysis were introduced in Part I. The first was to explain individual behavior in terms of individual and group variables. The second was to explain relationships in a single unit with context models of individual behavior. Details of procedures involving analyses of the first type are presented in this part.

Chapter 3 elaborates on the basic model of contextual analysis for continuous variables. The problem of interpreting effects in contextual analysis is addressed in Chapter 4, while Chapter 5 applies the model to categorical data. Extensions of the basic model of contextual analysis to more complex problems involving nonlinear context effects, multiple sets of variables, and multiple layers or types of groups are described in Chapters 6 and 7. Suggestions for computer assistance are contained in Appendix A.

Contextual Analysis for Continuous Individual-level Data

The basic model of contextual analysis described in Section 1.7 is examined more closely in this chapter.[1] This model specifies that an individual-level dependent variable is a function of a set of explanatory variables consisting of an individual-level variable, a group variable, and an interaction variable. The objective of this model is to detect the presence or absence of effects on the dependent variable associated with each of the three explanatory variables.

A hypothetical four-city study of social norms and deviance is used to illustrate concepts and procedures. In the analysis, the dependent variable Y is a measure of personal deviance and X is a measure of personal tolerance for deviance. The measure of Y is based on a continuous scale ranging from 0 to 100. The measure of X is based on a scale ranging from 0 to 10. Low values represent low incidents of deviance and low degrees of personal tolerance. The observed scores on these variables are presented in Table 3.1.

The sample of cities and the sample of individuals in each city are purposely small so that the discussion of concepts and procedures can easily be connected to actual data. Many of the operations described can be performed using a pocket calculator with the capability for simple regression. Related computer applications are contained in Appendix A, especially Illustrations 6 and 7.

3.1

The Conventional Approach Treating Only Individual-level Variables

The typical single-level analysis of a dependent individual-level variable for a general population was described in Section 1.2. A similar analysis is applied to the data in

[1] The references at the end of this book identify many works directly and indirectly related to contextual analysis. For a list of works more directly pertinent to this chapter, see footnote 1, Chapter 1. For references applying explicitly to methodological issues encountered in contextual analysis, see Chapter 13.

Table 3.1
Hypothetical Data and Analysis Summary Based on the Four-city Study for the First Example

CITY NUMBER k	DEVIANCE SCORE y_{ik}	TOLERANCE SCORE x_{ik}	CITY AVERAGES ON TOLERANCE $\bar{x}_k$	CITY REGRESSION LINES $y_{ik} = d_{0k} + d_{1k}x_{ik} + f_{ik}$
1	19	1	3	$y_{i1} = 14.0 + 2.5x_{i1} + f_{i1}$
	34	3	3	$R^2 = 0.095$
	4	3	3	
	29	5	3	
2	44	3	5	$y_{i2} = 26.0 + 6.0x_{i2} + f_{i2}$
	71	5	5	$R^2 = 0.390$
	41	5	5	
	68	7	5	
3	45	3	5	$y_{i3} = 30.5 + 4.0x_{i3} + f_{i3}$
	63	5	5	$R^2 = 0.211$
	33	5	5	
	61	7	5	
4	68	5	7	$y_{i4} = 28.0 + 7.5x_{i4} + f_{i4}$
	98	7	7	$R^2 = 0.353$
	58	7	7	
	98	9	7	

Separate equations

Intercept: $d_{0k} = b_0 + b_2\bar{x}_k + u_k$
$= 7.13 + 3.50\bar{x}_k + u_k, \quad R^2 = 0.610$

Slope: $d_{1k} = b_1 + b_3\bar{x}_k + v_k$
$= -1.25 + 1.25\bar{x}_k + v_k, \quad R^2 = 0.862$

Single equation

$y_{ik} = b_0 + b_1x_{ik} + b_2\bar{x}_k + b_3x_{ik}\bar{x}_k + e_{ik}$
$= -13.72 + 3.28x_{ik} + 8.03\bar{x}_k + 0.34x_{ik}\bar{x}_k + e_{ik}, \quad R^2 = 0.764$

Noncontext equation

$y_i = d_0 + d_1x_i + e_i$
$= 2.75 + 9.88x_i + e_i, \quad R^2 = 0.613$

Table 3.1. This involves the relationship between personal tolerance and deviance for all $n = 16$ cases, without regard for city differences. It might be theorized that persons who are generally tolerant of deviant behavior are more likely to engage in deviant behaviors than less tolerant individuals. The conventional way to study this relationship is with a simple regression model as in

$$y_i = d_0 + d_1x_i + e_i, \qquad i = 1,2, \ldots, n \tag{3.1}$$

In Eq. (3.1) y_i and x_i are the deviance and tolerance scores for the ith individual, and e_i represents the effect on Y of all variables other than X. The symbols d_0 and d_1 represent the regression intercept and slope, respectively. The values of the coefficients for this example are presented in Eq. (3.2):

$$y_i = 2.75 + 9.88x_i + e_i, \qquad R^2 = 0.613 \tag{3.2}$$

Because there are no variables referring to properties of groups, this equation is identified in Table 3.1 as the noncontext equation.

The numerical value of the intercept in this equation indicates that an individual having a score of 0.0 on the tolerance scale could be predicted, on the average, to have a score of 2.75 on the deviance scale. The coefficient for $X(9.88)$ indicates that for each unit change in tolerance, one would expect, on the average, an increase of about ten units in deviant behavior. The value of $R^2(0.613)$ indicates that 61.3 percent of the variance in deviance is explained by personal tolerance. From these results it normally would be inferred that personal tolerance is a strong predictor of deviance.

However, a conclusion that this analysis has revealed and measured a "true" individual-level relationship is open to question. It was simply assumed in this analysis that the relationships between Y and X in each of the cities identified in Table 3.1 differ only because of sampling fluctuation, and this may well not be true. There is a possibility that the cities differ significantly with respect to "climates," or levels of tolerance, and that these differences affect individual-level deviance independent of personal tolerance. In other words, some city context effect may be accounting for at least part of the apparent relationship observed at the individual level. Thus we need a framework to conceptualize the problem and a set of procedures to isolate individual and context effects.

3.2 *Recognizing Contextual Effects by Studying Within-group Relationships*

Contextual analysis requires the study of relationships *within* groups. Therefore the notation used in Eq. (3.1) is expanded to include subscript k identifying particular groups. In the four-city example, x_{ik} and y_{ik} denote the tolerance and deviance scores for the ith individual in the kth city. For example, $y_{31} = 4$ is the deviance score for the third individual in the first city. Similarly, d_{0k} and d_{1k} are the numerical values of the regression intercept and slope for the kth group. The symbol f_{ik} represents the effects of the residual variable for the ith individual in the kth group.

The regression model relating Y, X, and the residual variable within groups is expressed in the equation

$$y_{ik} = d_{0k} + d_{1k}x_{ik} + f_{ik} \tag{3.3}$$

Through a simple regression analysis in each of the k groups (cities), values of four intercepts and four slopes are obtained. Computer procedures are contained in Appendix A, Illustration 7. The results are presented in the right-hand side of Table 3.1. For example, the numerical values of the intercept and slope for city 1 are 14.00 and 2.50, respectively. The intercept in city 1 indicates that on the average an individual having a score of 0.0 on the tolerance scale could be predicted to have a score of 14 on the deviance scale. The numerical value of the slope indicates that for each unit increase in the tolerance score one would predict a change of 2.50 in the deviance score. The value of R^2 indicates that 9.5 percent of the variance in deviance scores is explained by the tolerance scores.

The most important observation which can be made about the values of the coefficients in Table 3.1 is that they differ substantially from city to city. For example, compare city 1 and city 4. The relationship between personal tolerance and deviance is affected by something associated with the city in which the individual resides. Contrary to expectation, individuals having the same score on personal tolerance but living in different city contexts, on the average, have different scores on deviance. It is also important to note that the relationship observed within a given city may be very different from the relationship observed for all the cases together. Compare the intercept and slope for city 1 to the intercept and slope obtained for all the cases considered together.

The differences between the relationships observed in the cities suggest the presence of some unidentified group effects on individual-level deviance. As observed in Chapter 1, the composite of these effects is detected using the analysis-of-covariance approach to explaining variation in an individual-level variable.[2] This approach detects the presence of community influences by introducing a nominal variable with categories for each community. However, the analysis-of-covariance approach does not identify specific group effects. Therefore in terms of theory development, the task is to connect the differences between the relationships observed in the cities to specific group variables.

Since the within-group regression coefficients d_{0k} and d_{1k} are measures of within-group relationships, the connections between the within-group relationships and specific properties of groups can be studied in the two equations

$$d_{0k} = f(w_k) + u_k \qquad (3.4)$$

$$d_{1k} = g(w_k) + v_k \qquad (3.5)$$

where w_k measures some property of the group. According to one's theory, this group property might account for the differences observed between the relationships in the four cities. In Eqs. (3.4) and (3.5), u_k and v_k stand for the effect on the intercepts and slopes of all group variables other than W. The variable W can be any group (city) property.[3] However, there is one property of a group that will frequently be of immediate concern. This variable, made up of the means of X for each city, is denoted by the symbol $\bar{X}$. Table 3.1 indicates that the means for the four groups are, respectively, 3, 5, 5, and 7.

The reason why the group mean variable is of special importance in a contextual analysis will become increasingly apparent. However, it is sufficient that the means frequently have a very direct interpretation as a context variable. In this instance it could have been theorized that the relationship between personal tolerance and deviance is different in cities with high *levels* of tolerance compared with cities with low levels. Thus the variable W in Eqs. (3.4) and (3.5) is taken to be $\bar{X}$. In this chapter

[2] The distinction between group effects detected in contextual analysis and group effects detected with analysis of covariance is discussed by Firebaugh (1977).

[3] For types of group variables, see Lazarsfeld and Menzel (1961); Riley (1964); Selvin and Hagstrom (1963); Eulau (1969); Cartwright (1969).

only the simple linear function is considered. Nonlinear functions are discussed in Chapter 6. Therefore according to the specifications of the basic model of contextual analysis, we have for the within-group slopes and intercepts

$$d_{0k} = b_0 + b_2\bar{x}_k + u_k \tag{3.6}$$

$$d_{1k} = b_1 + b_3\bar{x}_k + v_k \tag{3.7}$$

where the coefficients indicate the presence of individual and context effects. Where group sizes n_k vary substantially, estimates of the coefficients in these equations should be obtained from a weighted regression procedure. Details of this procedure are contained in Appendix B.

In the following sections the implications of this formulation of contextual analysis are drawn out for all logical possibilities. The purpose of examining each is to show how effects are manifest. It should not be assumed that the occurrence of each is equally probable.

To illustrate the several models, the hypothetical data for the four-city study are made to fit each case. The computations for the examples can be done by hand or with the assistance of a pocket calculator with the capability for simple regression. Related computer procedures are contained in Appendix A, Illustration 7.

3.2.1. Individual-level Effect Only

When the explanatory variable X affects the dependent variable Y only at the level of the individual, the regression line of Y on X is the same in each group. When the lines are identical and the common slope is not equal to zero, two people with the same value of X will, on the average, have the same value of Y whether they belong to the same or different groups. Thus there is no specified group mean effect. That is, there is no group effect directly associated with the group means.[4] On the other hand, two individuals belonging to the same group but having different values of X will, on the average, have different values of Y. Thus there is an individual-level effect of X.

This can be formally expressed by denoting the common intercept of these lines as b_0 and their common slope as b_1. The intercept in the kth group, d_{0k}, equals b_0 except for a residual term u_k. Similarly, the slope d_{1k} equals b_1 except for a residual term v_k. These residuals measure effects of other group-level variables which have not been explicitly measured in the analysis. For example, cities might differ in terms of racial composition or law enforcement, and these factors might influence individual-level deviance. These statements are formally expressed in the two equations

$$d_{0k} = b_0 + u_k \tag{3.8}$$

$$d_{1k} = b_1 + v_k \tag{3.9}$$

[4] This does not imply that there are no group effects in this situation. It can generally be assumed that there are unspecified group influences. These unspecified group effects will generally emerge in an analysis of covariance of the data (cf. Firebaugh 1977).

An equivalent way to visualize the case of individual effects only is in terms of Eqs. (3.6) and (3.7) where the coefficients b_2 and b_3 equal zero, indicating that the intercepts and slopes do not depend on the group means.

Data conforming to the individual effect only for the four-city study are given in Table 3.2. This model is illustrated graphically in Figure 3.1 where the fitted regression lines for the four cities are shown without residuals. Figure 3.1(a) shows the within-group (city) regression lines. Since their slopes and intercepts are identical, they produce a single straight line. The graph of Figure 3.1(b) shows that the slopes and the intercepts do not vary with the contextual variable $\bar{X}$. Table 3.2 indicates that the numerical values of b_0 and b_1 are 25.00 and 6.50, respectively, for this example.

Table 3.2
Hypothetical Data and Analysis Summary Based on the Four-city Study for an Individual-level Effect Only

CITY NUMBER k	DEVIANCE SCORE y_{ik}	TOLERANCE SCORE x_{ik}	CITY AVERAGES ON TOLERANCE $\bar{x}_k$	CITY REGRESSION LINES $y_{ik} = d_{0k} + d_{1k}x_{ik} + f_{ik}$
1	25	0	2	$y_{i1} = 25.0 + 6.5x_{i1} + f_{i1}$
	53	2	2	$R^2 = 0.430$
	23	2	2	
	51	4	2	
2	38	2	4	$y_{i2} = 25.0 + 6.5x_{i2} + f_{i2}$
	66	4	4	$R^2 = 0.430$
	36	4	4	
	64	6	4	
3	51	4	6	$y_{i3} = 25.0 + 6.5x_{i3} + f_{i3}$
	79	6	6	$R^2 = 0.430$
	49	6	6	
	77	8	6	
4	64	6	8	$y_{i4} = 25.0 + 6.5x_{i4} + f_{i4}$
	92	8	8	$R^2 = 0.430$
	62	8	8	
	90	10	8	

Separate equations

Intercept: $d_{0k} = b_0 + u_k = 25.0 + u_k$

Slope: $d_{1k} = b_1 + v_k = 6.5 + v_k$

Single equation

$$y_{ik} = b_0 + b_1x_{ik} + b_2\bar{x}_k + b_3x_{ik}\bar{x}_k + e_{ik}$$
$$= 25.0 + 6.5x_{ik} + 0.0\bar{x}_k + 0.0x_{ik}\bar{x}_k + e_{ik}, \qquad R^2 = 0.724$$

Noncontext equation

$$y_i = d_0 + d_1x_i + e_i$$
$$= 25.0 + 6.5x_i + e_i, \qquad R^2 = 0.851$$

Figure 3.1
Individual Effect Only (Residuals not Shown)

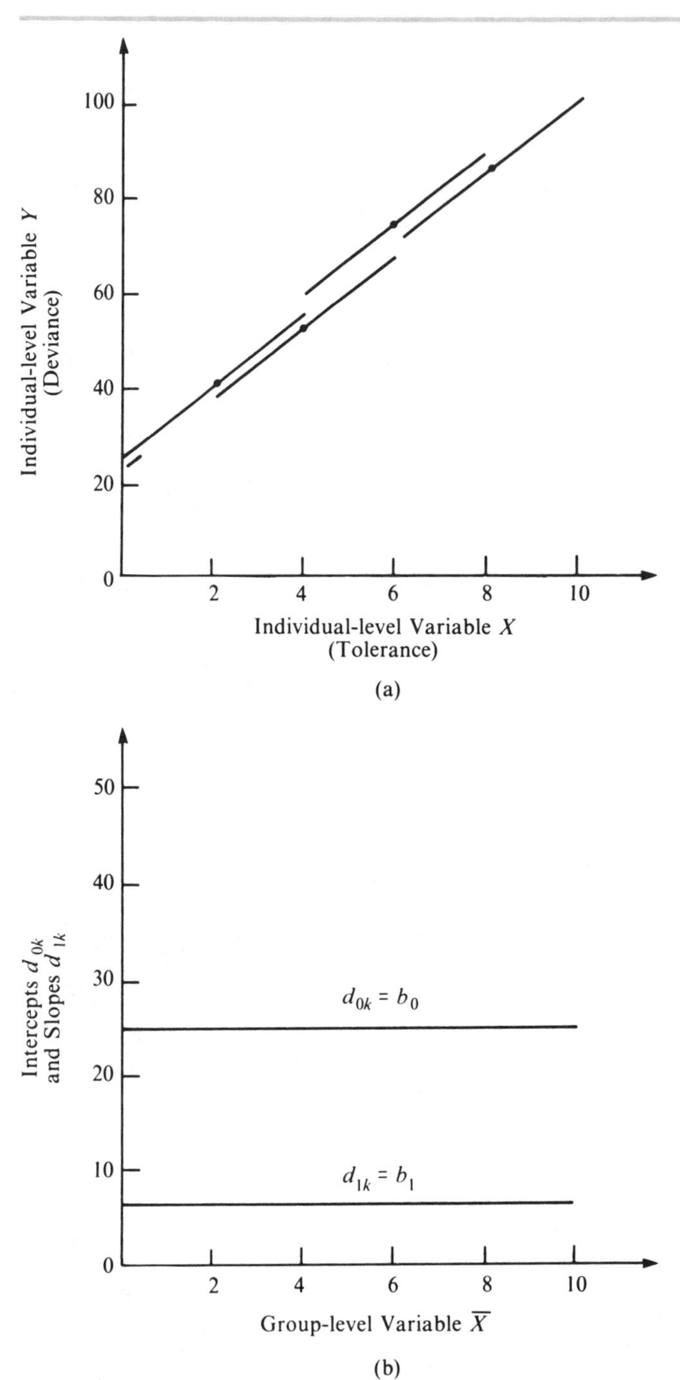

3.2.2. Group-level Effect Only

In the presence of a group effect only, there is no relationship between Y and X in any of the groups. Therefore the regression lines for Y on X are horizontal and the slopes d_{1k} equal zero. However, the intercepts d_{0k} depend upon the contextual variable $\bar{X}$ and a residual term. These relationships are expressed in the following equations:

$$d_{0k} = b_0 + b_2\bar{x}_k + u_k \tag{3.10}$$

$$d_{1k} = v_k \tag{3.11}$$

In terms of Eq. (3.7), the coefficients b_1 and b_3 equal zero, indicating that the slopes are constant and equal to zero except for the residual term.

Data conforming to the case for a group effect only are given in Table 3.3, and

Table 3.3
Hypothetical Data and Analysis Summary Based on the Four-city Study for a Group-level Effect Only

CITY NUMBER k	DEVIANCE SCORE y_{ik}	TOLERANCE SCORE x_{ik}	CITY AVERAGES ON TOLERANCE $\bar{x}_k$	CITY REGRESSION LINES $y_{ik} = d_{0k} + d_{1k}x_{ik} + f_{ik}$
1	23	0	2	$y_{i1} = 23.0 + 0.0x_{i1} + f_{i1}$
	38	2	2	$R^2 = 0$
	8	2	2	
	23	4	2	
2	36	2	4	$y_{i2} = 36.0 + 0.0x_{i2} + f_{i2}$
	51	4	4	$R^2 = 0$
	21	4	4	
	36	6	4	
3	49	4	6	$y_{i3} = 49.0 + 0.0x_{i3} + f_{i3}$
	64	6	6	$R^2 = 0$
	34	6	6	
	49	8	6	
4	62	6	8	$y_{i4} = 62.0 + 0.0x_{i4} + f_{i4}$
	77	8	8	$R^2 = 0$
	47	8	8	
	62	10	8	

Separate equations

Intercept: $d_{0k} = b_0 + b_2\bar{x}_k + u_k$
$= 10.0 + 6.5\bar{x}_k + u_k$

Slope: $d_{1k} = b_1 + v_k$
$= 0.0 + v_k$

Single equation

$y_{ik} = b_0 + b_1x_{ik} + b_2\bar{x}_k + b_3x_{ik}\bar{x}_k + e_{ik}$
$= 10.0 + 0.0x_{ik} + 6.5\bar{x}_k + 0.0x_{ik}\bar{x}_k + e_{ik}$, $R^2 = 0.652$

Noncontext equation

$y_i = d_0 + d_1x_i + e_i$
$= 19.29 + 4.64x_i + e_i$, $R^2 = 0.466$

Figure 3.2
Group Effect Only (Residuals not Shown)

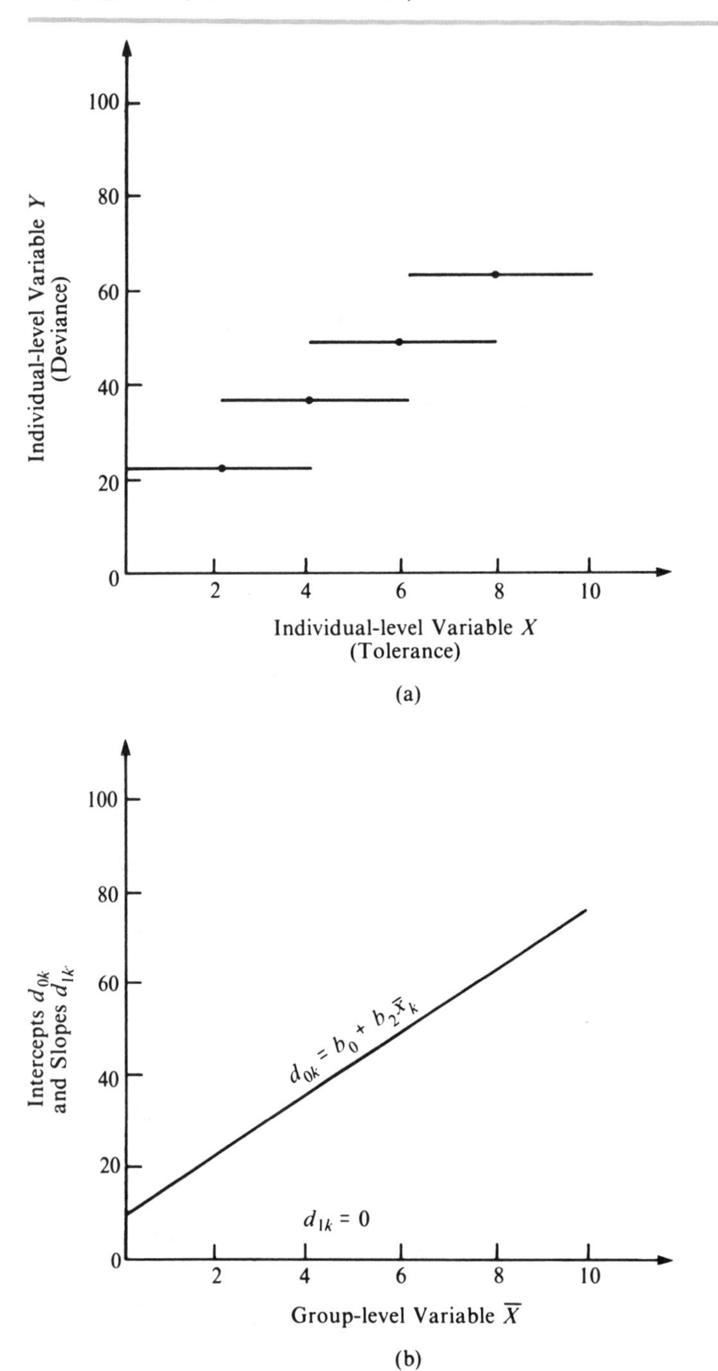

the case is graphically illustrated in Figure 3.2. The absence of the individual effect is seen from the regression lines of Y on X with zero slopes. The presence of the group effect is indicated in Figure 3.2(a) by the fact that the within-group intercepts are different.

Figure 3.2(b) indicates the presence of a group effect in the nonzero slope of the line relating the within-group intercepts and the group means. Table 3.3 indicates that the numerical value of this slope is 6.50. The zero intercept and slope relating the group slopes and the means indicate the absence of individual and interaction effects.

3.2.3. Interaction Effect Only

In the case of an interaction effect only, the within-group intercepts d_{0k} are constant, indicating the absence of a pure group effect. The coefficient b_2 in Eq. (3.6) therefore equals zero. In addition, the coefficient b_1 in Eq. (3.7) is also equal to zero, indicating the absence of a pure individual-level effect. However, when there is an interaction effect involving individual- and group-mean effects, the within-group slopes are not constant as in the previous cases. They depend on the context variable $\bar{X}$ and a residual variable. These relationships are shown in the following equations:

$$d_{0k} = b_0 + u_k \tag{3.12}$$

$$d_{1k} = b_3\bar{x}_k + v_k \tag{3.13}$$

Data conforming to this model for the four-city study are given in Table 3.4, and the case is illustrated in Figure 3.3. In Figure 3.3(a) it can be observed in city 4, where the tolerance level is high, that two individuals with different tolerance scores X will, on the average, differ substantially on their deviance scores Y. However, this difference decreases with the tolerance levels in the other cities and vanishes at the theoretical point where $\bar{X}$ equals zero. Similarly, two individuals with the same tolerance score but residing in different cities will, on the average, have different deviance scores. But this difference also decreases with diminishing tolerance levels and vanishes at the theoretical point where $\bar{X}$ equals zero. In the case of an interaction effect only, the individual- and group-level variables do have an effect on the dependent variable, but *only* in combination. Since this interaction phenomenon involves variables measured at different levels of observation, it is referred to as a "cross-level interaction."

In Figure 3.3(a) the presence of an interaction effect is observed in the different within-group slopes. Figure 3.3(b) reflects the presence of an interaction effect by the nonzero slope of the line relating the within-group slopes to the group means. Table 3.4 indicates that the numerical value of this slope is 0.75. In this example the value of the slope is small relative to the other coefficients because of the differences in the ranges of the variables d_{0k}, d_{1k}, and the group means $\bar{X}$.

Table 3.4
Hypothetical Data and Analysis Summary Based on the Four-city Study for an Interaction Effect Only

CITY NUMBER k	DEVIANCE SCORE y_{ik}	TOLERANCE SCORE x_{ik}	CITY AVERAGES ON TOLERANCE $\bar{x}_k$	CITY REGRESSION LINES $y_{ik} = d_{ok} + d_{1k}x_{ik} + f_{ik}$
1	20	0	2	$y_{i1} = 20.0 + 1.5x_{i1} + f_{i1}$
	38	2	2	$R^2 = 0.038$
	8	2	2	
	26	4	2	
2	26	2	4	$y_{i2} = 20.0 + 3.0x_{i2} + f_{i2}$
	47	4	4	$R^2 = 0.138$
	17	4	4	
	38	6	4	
3	38	4	6	$y_{i3} = 20.0 + 4.5x_{i3} + f_{i3}$
	62	6	6	$R^2 = 0.265$
	32	6	6	
	56	8	6	
4	56	6	8	$y_{i4} = 20.0 + 6.0x_{i4} + f_{i4}$
	83	8	8	$R^2 = 0.390$
	53	8	8	
	80	10	8	

Separate equations

Intercept: $d_{ok} = b_0 + u_k$
$= 20.0 + u_k$

Slope: $d_{1k} = b_3\bar{x}_k + v_k$
$= 0.75\bar{x}_k + v_k$

Single equation

$$y_{ik} = b_0 + b_1x_{ik} + b_2\bar{x}_k + b_3x_{ik}\bar{x}_k + e_{ik}$$
$$= 20.0 + 0.0x_{ik} + 0.0\bar{x}_k + 0.75x_{ik}\bar{x}_k + e_{ik}, \quad R^2 = 0.742$$

Noncontext equation

$$y_i = d_0 + d_1x_i + e_i$$
$$= 10.36 + 6.43x_i + e_i, \quad R^2 = 0.663$$

3.2.4. Individual and Group Effects

When there is an individual effect, the relationship between Y and X within the groups is characterized by all the nonzero slopes being the same and equal to b_1 except for a residual term. In the presence of a specified group effect, associated with the means of X, the intercepts depend upon $\bar{X}$ and a residual variable. These relationships are expressed in the following equations:

$$d_{ok} = b_0 + b_2\bar{x}_k + u_k \tag{3.14}$$

Figure 3.3
Interaction Effect Only (Residuals not Shown)

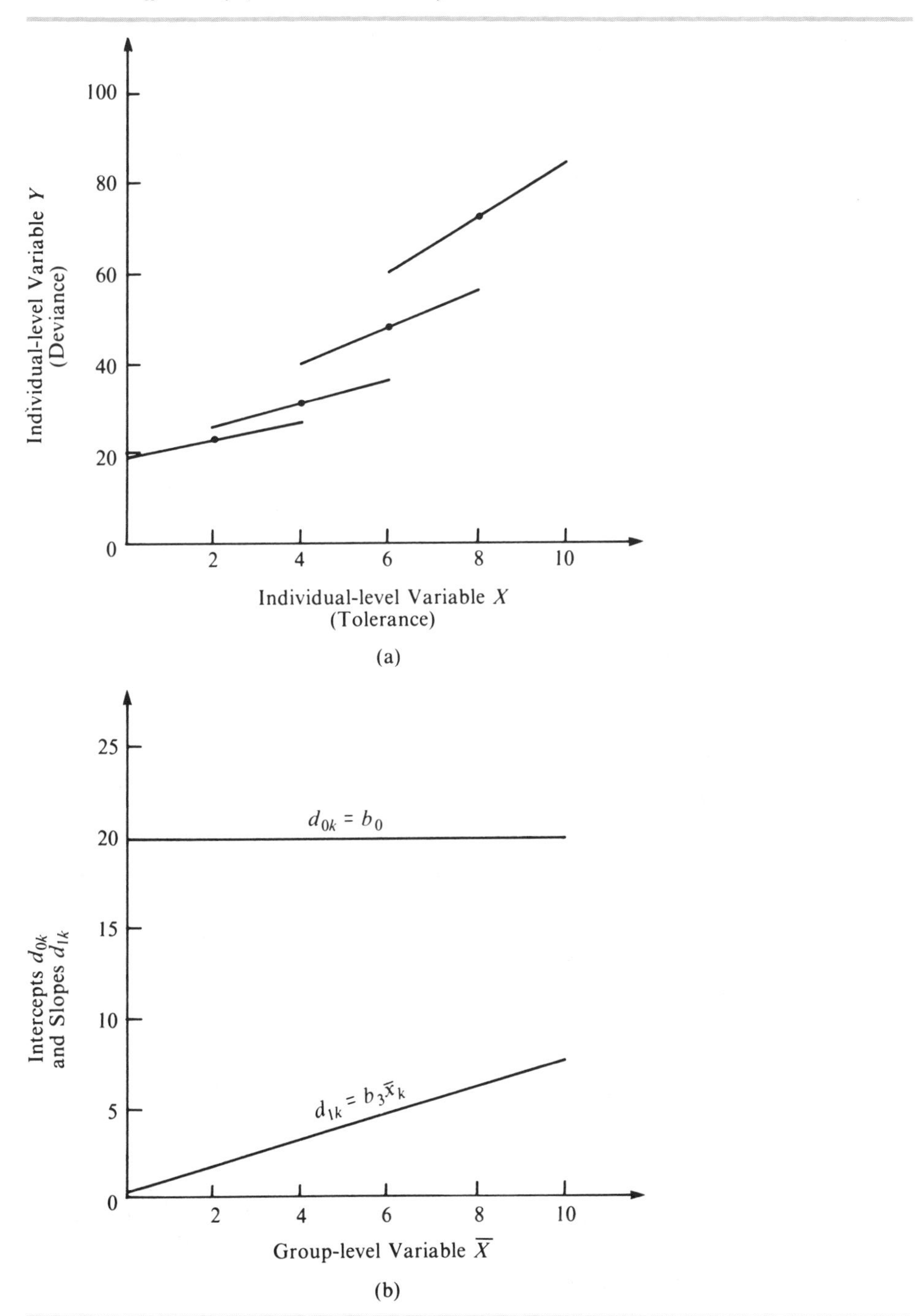

$$d_{1k} = b_1 + v_k \tag{3.15}$$

Data conforming to the case of individual and group effects for the four-city study are given in Table 3.5, and the case is illustrated graphically in Figure 3.4. Figure 3.4(a) shows the relationship between Y and X for each city. In Table 3.3 and in Figure 3.2 the nonzero slopes of the regression lines for each city indicate the presence of the individual-level effect of personal tolerance. Within a given city, two individuals with different tolerance scores X will, on the average, have different deviance scores Y. Thus personal tolerance affects deviance. The effect does not vary across groups, indicating the absence of a cross-level interaction.

Table 3.5
Hypothetical Data and Analysis Summary Based on the Four-city Study for Individual- and Group-level Effects

CITY NUMBER k	DEVIANCE SCORE y_{ik}	TOLERANCE SCORE x_{ik}	CITY AVERAGES ON TOLERANCE $\bar{x}_k$	CITY REGRESSION LINES $y_{ik} = d_{0k} + d_{1k}x_{ik} + f_{ik}$
1	10	0	2	$y_{i1} = 10.0 + 5.5x_{i1} + f_{i1}$
	36	2	2	$R^2 = 0.350$
	6	2	2	
	32	4	2	
2	31	2	4	$y_{i2} = 20.0 + 5.5x_{i2} + f_{i2}$
	57	4	4	$R^2 = 0.350$
	27	4	4	
	53	6	4	
3	52	4	6	$y_{i3} = 30.0 + 5.5x_{i3} + f_{i3}$
	78	6	6	$R^2 = 0.350$
	48	6	6	
	74	8	6	
4	73	6	8	$y_{i4} = 40.0 + 5.5x_{i4} + f_{i4}$
	99	8	8	$R^2 = 0.350$
	69	8	8	
	95	10	8	

Separate equations

Intercept: $d_{0k} = b_0 + b_2\bar{x}_k + u_k$
$= 0.0 + 5.0\bar{x}_k + u_k$

Slope: $d_{1k} = b_1 + v_k$
$= 5.5 + v_k$

Single equation

$y_{ik} = b_0 + b_1x_{ik} + b_2\bar{x}_k + b_3x_{ik}\bar{x}_k + e_{ik}$
$= 0.0 + 5.5x_{ik} + 5.0\bar{x}_k + 0.0x_{ik}\bar{x}_k + e_{ik},\quad R^2 = 0.845$

Noncontext equation

$y_i = d_0 + d_1x_i + e_i$
$= 7.14 + 9.07x_i + e_i,\quad R^2 = 0.795$

Figure 3.4
Individual and Group Effects (Residuals not Shown)

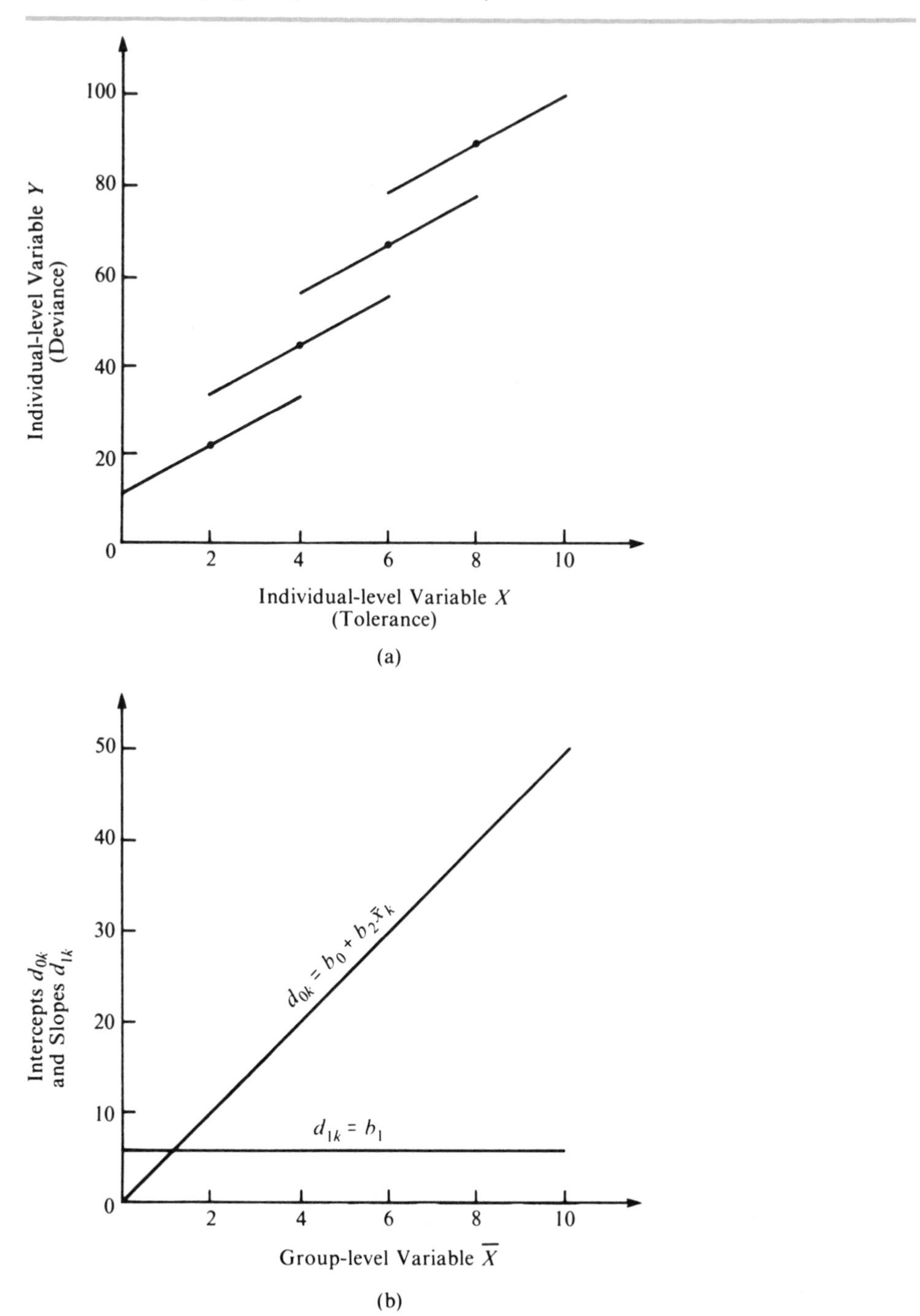

Similarly, the different intercepts of the regression lines show the presence of the group effect. Two individuals with the same tolerance score but residing in different cities will, on the average, have different deviance scores. The amount of this difference does not vary across groups, indicating the absence of a cross-level interaction effect. Figure 3.4(b) shows how the slopes and intercepts of the regression lines within the groups are related to the contextual variable. The intercepts vary with $\bar{X}$. In this example they increase with $\bar{X}$. The slopes are constant and are not affected by $\bar{X}$. Table 3.5 shows that the numerical values of b_1 and b_2 are 5.50 and 5.00, respectively.

Table 3.6
Hypothetical Data and Analysis Summary Based on the Four-city Study for Individual and Interaction Effects

CITY NUMBER k	DEVIANCE SCORE y_{ik}	TOLERANCE SCORE x_{ik}	CITY AVERAGES ON TOLERANCE $\bar{x}_k$	CITY REGRESSION LINES $y_{ik} = d_{0k} + d_{1k}x_{ik} + f_{ik}$
1	10	0	2	$y_{i1} = 10.0 + 5.5x_{i1} + f_{i1}$
	36	2	2	$R^2 = 0.349$
	6	2	2	
	32	4	2	
2	23	2	4	$y_{i2} = 10.0 + 6.5x_{i2} + f_{i2}$
	51	4	4	$R^2 = 0.429$
	21	4	4	
	49	6	4	
3	40	4	6	$y_{i3} = 10.0 + 7.5x_{i3} + f_{i3}$
	70	6	6	$R^2 = 0.500$
	40	6	6	
	70	8	6	
4	61	6	8	$y_{i4} = 10.0 + 8.5x_{i4} + f_{i4}$
	93	8	8	$R^2 = 0.560$
	63	8	8	
	95	10	8	

Separate equations

Intercept: $d_{0k} = b_0 + u_k$
$= 10.0 + u_k$

Slope: $d_{1k} = b_1 + b_3\bar{x}_k + v_k$
$= 4.5 + 0.5\bar{x}_k + v_k$

Single equation

$$y_{ik} = b_0 + b_1x_{ik} + b_2\bar{x}_k + b_3x_{ik}\bar{x}_k + e_{ik}$$
$$= 10.0 + 4.5x_{ik} + 0.0\bar{x}_k + 0.5x_{ik}\bar{x}_k + e_{ik}, \qquad R^2 = 0.832$$

Noncontext equation

$$y_i = d_0 + d_1x_i + e_i$$
$$= 3.57 + 8.79x_i + e_i, \qquad R^2 = 0.809$$

3.2.5. Individual and Interaction Effects

In this case the coefficients b_1 and b_3 are not equal to zero, indicating the presence of individual- and cross-level interaction effects. The coefficient b_2 equals zero, indicating the absence of a pure group effect. Thus for this special case, the equations are

$$d_{0k} = b_0 + u_k \tag{3.16}$$

$$d_{1k} = b_1 + b_3\bar{x}_k + v_k \tag{3.17}$$

Data conforming to the model are given in Table 3.6, and the case is illustrated in Figure 3.5. All the lines converge at a common intercept, indicating the absence of a pure group effect. The difference between this case and the case involving an interaction effect only (Figure 3.3) is that the slopes do not become zero at the theoretical point where $\bar{X}$ equals zero. According to this model, personal tolerance does affect deviant behavior independent of the city norms involving deviance. But this effect is strongest in cities most accepting of deviance. The numerical values of b_1 and b_3 in this example are 4.50 and 0.50, respectively.

3.2.6. Group and Interaction Effects

For this special case the coefficient b_1 equals zero, indicating the absence of pure individual-level effects. Equations for the model are

$$d_{0k} = b_0 + b_2\bar{x}_k + u_k \tag{3.18}$$

$$d_{1k} = b_3\bar{x}_k + v_k \tag{3.19}$$

Data consistent with this case are given in Table 3.7, and it is illustrated in Figure 3.6. When group and interaction effects exist, personal tolerance of deviance has no effect on deviant behavior except where there is at least some degree of tolerance in the normative climate. The numerical values of b_2 and b_3 are 4.50 and 0.50.

3.2.7. Individual, Group, and Interaction Effects

When all three effects are involved, the coefficients corresponding to individual, group, and interaction effects are different from zero. The equations for this model are

$$d_{0k} = b_0 + b_2\bar{x}_k + u_k \tag{3.20}$$

$$d_{1k} = b_1 + b_3\bar{x}_k + v_k \tag{3.21}$$

In subsequent discussion, this model is referred to as the "basic model of contextual analysis." This distinguishes it from model extensions treated later. The procedure

Figure 3.5
Individual and Interaction Effects (Residuals not Shown)

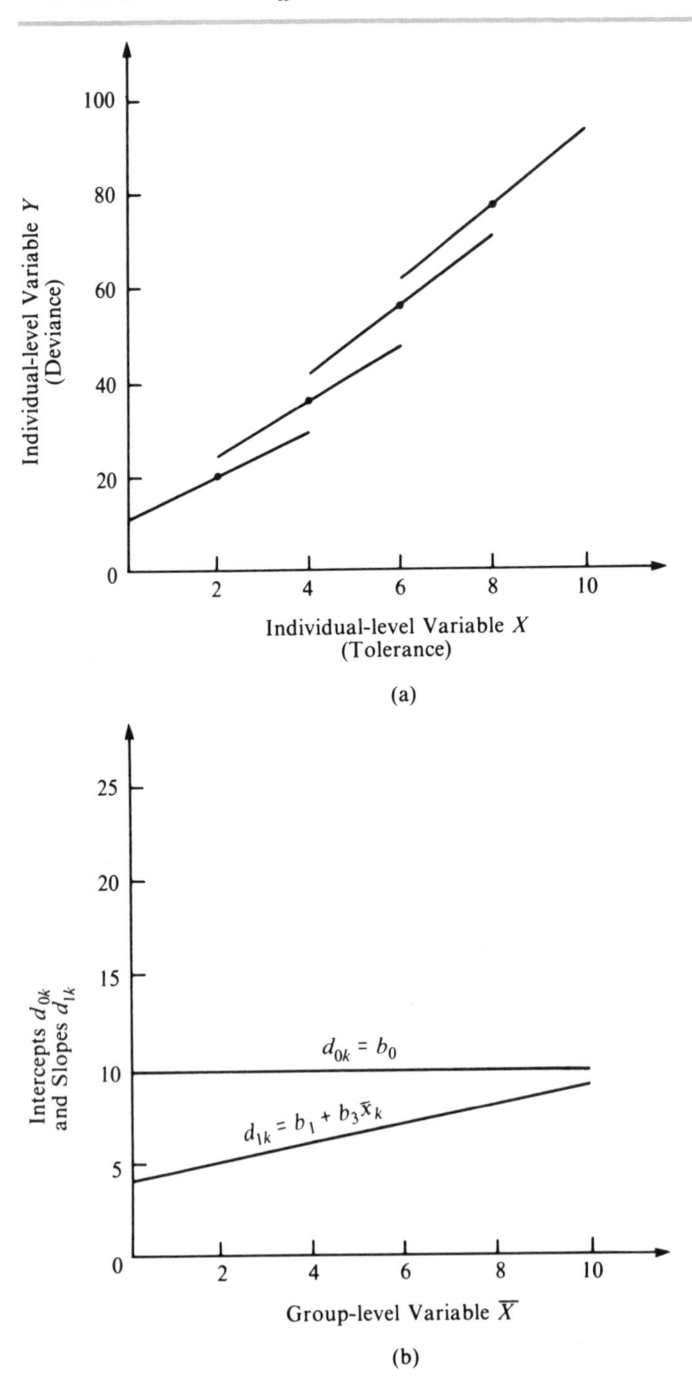

Table 3.7
Hypothetical Data and Analysis Summary Based on the Four-city Study for Group and Interaction Effects

CITY NUMBER k	DEVIANCE SCORE y_{ik}	TOLERANCE SCORE x_{ik}	CITY AVERAGES ON TOLERANCE $\bar{x}_k$	CITY REGRESSION LINES $y_{ik} = d_{0k} + d_{1k}x_{ik} + f_{ik}$
1	15	0	2	$y_{i1} = 15.0 + 1.0x_{i1} + f_{i1}$
	32	2	2	$R^2 = 0.017$
	2	2	2	
	19	4	2	
2	28	2	4	$y_{i2} = 24.0 + 2.0x_{i2} + f_{i2}$
	47	4	4	$R^2 = 0.066$
	17	4	4	
	36	6	4	
3	45	4	6	$y_{i3} = 33.0 + 3.0x_{i3} + f_{i3}$
	66	6	6	$R^2 = 0.138$
	36	6	6	
	57	8	6	
4	66	6	8	$y_{i4} = 42.0 + 4.0x_{i4} + f_{i4}$
	89	8	8	$R^2 = 0.222$
	59	8	8	
	82	10	8	

Separate equations

Intercept: $d_{0k} = b_0 + b_2\bar{x}_k + u_k$
$= 6.0 + 4.5\bar{x}_k + u_k$

Slope: $d_{1k} = b_3\bar{x}_k + v_k$
$= 0.5\bar{x}_k + v_k$

Single equation

$$y_{ik} = b_0 + b_1x_{ik} + b_2\bar{x}_k + b_3x_{ik}\bar{x}_k + e_{ik}$$
$$= 6.0 + 0.0x_{ik} + 4.5\bar{x}_k + 0.5x_{ik}\bar{x}_k + e_{ik}, \qquad R^2 = 0.807$$

Noncontext equation

$$y_i = d_0 + d_1x_i + e_i$$
$$= 6.0 + 7.5x_i + e_i, \qquad R^2 = 0.676$$

for relating the regression coefficients to the group means is referred to as the "separate-equation approach" to distinguish it from the "single-equation approach" below.

In terms of the example, it would be found that personal tolerance of deviance affects deviant behavior independent of the normative climate, the normative climate directly affects personal deviance, and the normative climate affects the degree to which personal tolerance affects deviance. Data conforming to the basic model are given in Table 3.8, and the model is presented visually in Figure 3.7. Table 3.8 shows that the numerical estimates of the coefficients b_0, b_1, b_2, and b_3 are 7.50, 1.00, 3.00, and 0.75, respectively.

Figure 3.6
Group and Interaction Effects (Residuals not Shown)

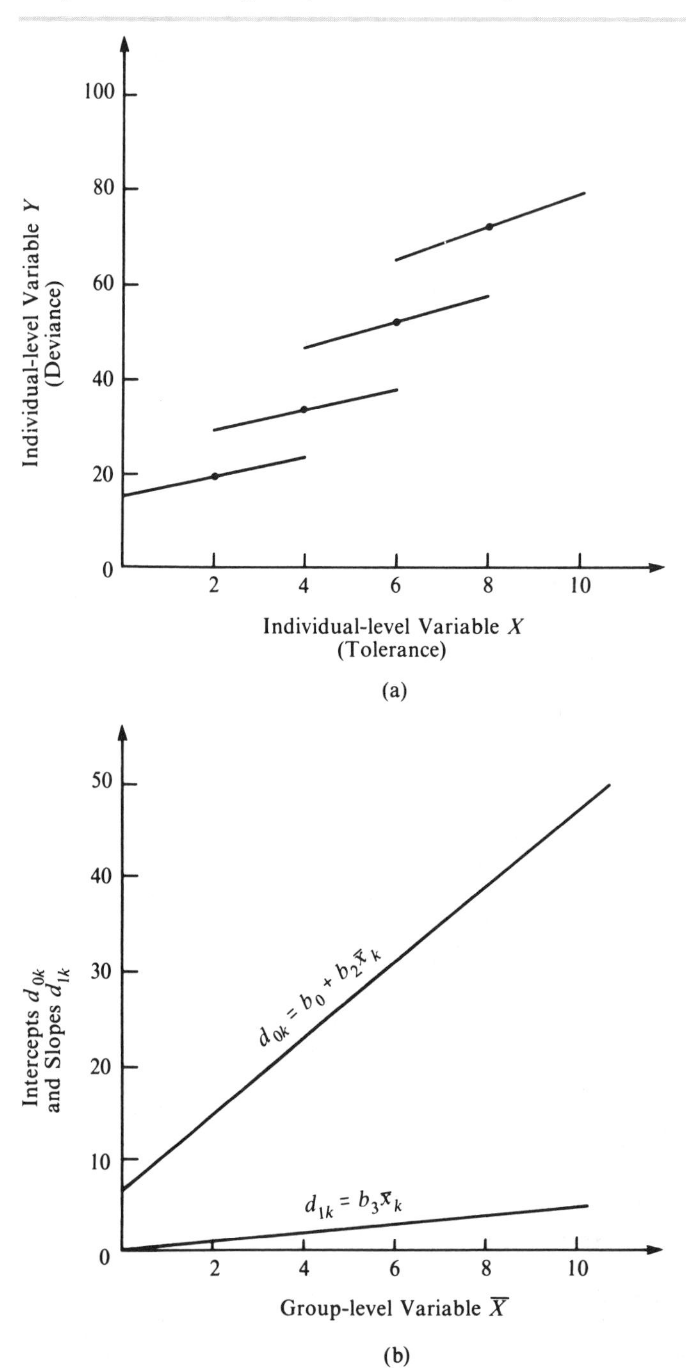

Table 3.8
Hypothetical Data and Analysis Summary Based on the Four-city Study for Individual, Group, and Interaction Effects

CITY NUMBER k	DEVIANCE SCORE y_{ik}	TOLERANCE SCORE x_{ik}	CITY AVERAGES ON TOLERANCE $\bar{x}_k$	CITY REGRESSION LINES $y_{ik} = d_{0k} + d_{1k}x_{ik} + f_{ik}$
1	10	0	2	$y_{i1} = 10.0 + 2.5x_{i1} + f_{i1}$
	30	2	2	$R^2 = 0.100$
	0	2	2	
	20	4	2	
2	24	2	4	$y_{i2} = 16.0 + 4.0x_{i2} + f_{i2}$
	47	4	4	$R^2 = 0.221$
	17	4	4	
	40	6	4	
3	44	4	6	$y_{i3} = 22.0 + 5.5x_{i3} + f_{i3}$
	70	6	6	$R^2 = 0.349$
	40	6	6	
	66	8	6	
4	70	6	8	$y_{i4} = 28.0 + 7.0x_{i4} + f_{i4}$
	99	8	8	$R^2 = 0.465$
	69	8	8	
	98	10	8	

Separate equations

Intercept: $d_{0k} = b_0 + b_2\bar{x}_k + u_k$
$= 4.00 + 3.00\bar{x}_k + u_k$

Slope: $d_{1k} = b_1 + b_3\bar{x}_k + v_k$
$= 1.0 + 0.75\bar{x}_k + v_k$

Single equation

$$y_{ik} = b_0 + b_1x_{ik} + b_2\bar{x}_k + b_3x_{ik}\bar{x}_k + e_{ik}$$
$$= 4.00 + 1.0x_{ik} + 3.00\bar{x}_k + 0.75x_{ik}\bar{x}_k + e_{ik}, \qquad R^2 = 0.865$$

Noncontext equation

$$y_i = d_0 + d_1x_i + e_i$$
$$= -1.35 + 9.57x_i + e_i, \qquad R^2 = 0.769$$

3.3
The Single-equation Approach to Contextual Analysis

The preceding conceptualization of individual, group, and interaction effects involves a direct focus on the relationships of the within-group regression intercepts and slopes to the within-group means. An important alternative is to estimate the various

Figure 3.7
Individual, Group, and Interaction Effects (Residuals not Shown)

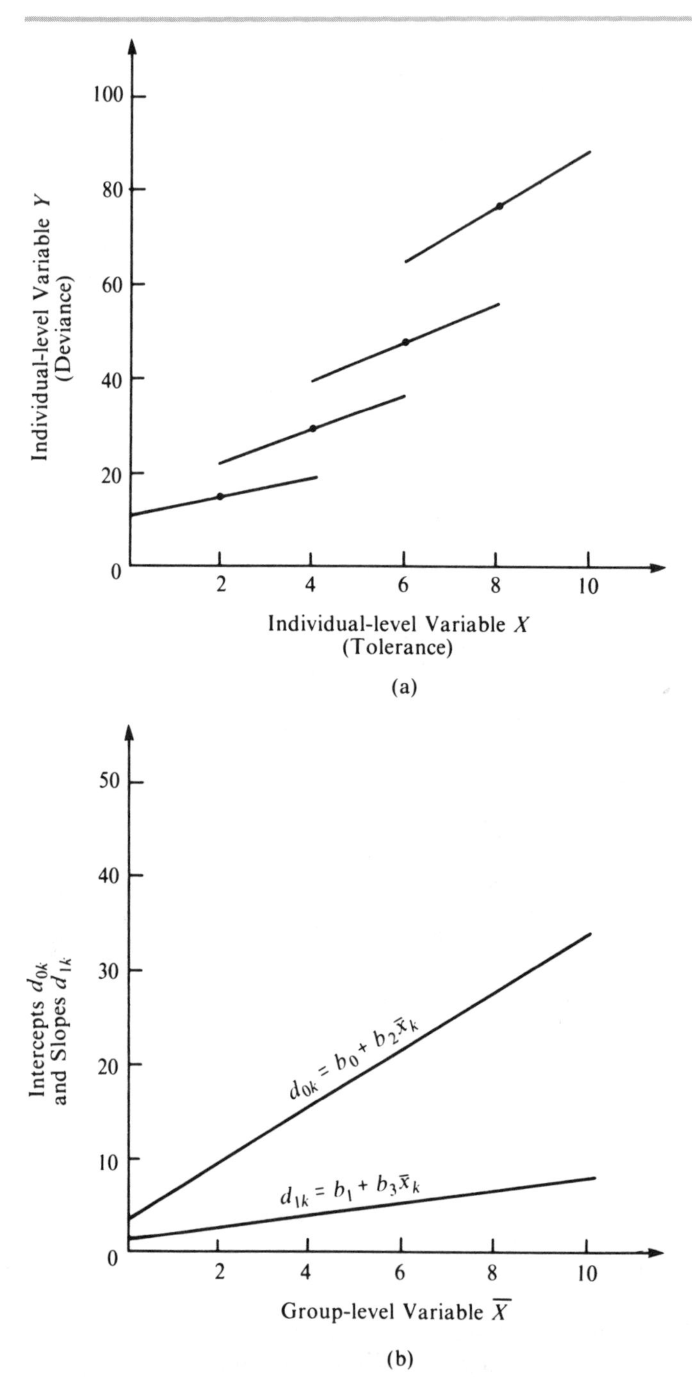

effects directly through a single multiple-regression analysis. Do this by substituting Eqs. (3.6) and (3.7) for d_{0k} and d_{1k} in Eq. (3.3). This produces the basic contextual model equation

$$y_{ik} = (b_0 + b_2\bar{x}_k + u_k) + (b_1 + b_3\bar{x}_k + v_k)x_{ik} + f_{ik} \tag{3.22}$$

Expanding this gives

$$y_{ik} = b_0 + b_2\bar{x}_k + u_k + b_1 x_{ik} + b_3\bar{x}_k x_{ik} + v_k x_{ik} + f_{ik} \tag{3.23}$$

Collecting the terms containing residuals gives

$$y_{ik} = b_0 + b_1 x_{ik} + b_2\bar{x}_k + b_3 x_{ik}\bar{x}_k + (u_k + v_k x_{ik} + f_{ik}) \tag{3.24}$$

If the last three terms in this equation are considered to be equal to e_{ik}, this equation can be written

$$y_{ik} = b_0 + b_1 x_{ik} + b_2\bar{x}_k + b_3 x_{ik}\bar{x}_k + e_{ik} \tag{3.25}$$

In the separate equation procedure, the means of X were used to construct an aggregate variable consisting of one observation for each community. The difference in the single-equation procedure is that the means of X are assigned to each of the individuals in the groups. Thus all individuals in city 1 get a value of 2 on the group-mean variable. All individuals in city 2 get a 4, and so on until every individual has a score on this variable, and the number of observations for this variable is the same as for the individual-level variable. The interaction variable is simply the product of these two variables. Computer procedures for constructing the group-level variables and for estimating the coefficients in single-equation models of contextual analysis are contained in Appendix A, Illustration 6.

The residuals e_{ik}, u_k, $v_k x_{ik}$, and f_{ik} are related according to the following equation:

$$e_{ik} = u_k + v_k x_{ik} + f_{ik} \tag{3.26}$$

Because e_{ik} is partly determined by x_{ik}, the residual e_{ik} in Eq. (3.25) may be correlated with the three explanatory variables. Conditions under which estimates in Eq. (3.25) equal the estimates obtained from the separate equations are discussed in Appendix C. Except for the data set given in Table 3.1, all the above data sets are made to have the two methods produce the same estimates. The reason for the different estimates in the first example is examined in detail in Chapter 6 under the discussion of nonlinear-context effects.

The partial regression coefficients b_1, b_2, and b_3 in Eq. (3.24) have the same interpretation in terms of multilevel effects as their counterparts in the separate-equation approach. A nonzero value of b_1 indicates the presence of an individual-level effect. A nonzero value of b_2 indicates the presence of a group effect. A nonzero value of b_3 indicates the presence of a cross-level interaction effect. A desirable characteristic of the single-equation procedure is that it makes explicit the idea of evaluating one level of effects while controlling for the effects from other levels. For example, in the last four-city example considered above, single-equation estimates for the basic model are

$$y_{ik} = 4.00 + 1.00x_{ik} + 3.00\bar{x}_k + 0.75x_{ik}\bar{x}_k + e_{ik} \tag{3.27}$$

In this equation, $b_1 = 1.00$ is an estimate of the personal effects of tolerant attitudes toward deviance while controlling for the contextual effects of city norms regarding deviance and the cross-level interaction effects involving personal tolerance and levels of tolerance.

Given the single-equation approach, it is possible to visualize all the models described in Section 3.2 as special cases of the basic model. For example, the case of individual effects only is observed when the coefficients b_2 and b_3 equal zero and b_1 does not equal zero. The single equations and estimates for each case described above are contained in Tables 3.2 through 3.8. The resulting typology of cases is summarized in Table 3.9.

Table 3.9
Typology of Contextual Effects when Studying Within-group Relationships

MODEL	INDIVIDUAL EFFECTS b_1	GROUP EFFECTS b_2	INTERACTION EFFECTS b_3
Individual effects only	$\neq 0$	$= 0$	$= 0$
Group effects only	$= 0$	$\neq 0$	$= 0$
Interaction effect only	$= 0$	$= 0$	$\neq 0$
Individual and group effects	$\neq 0$	$\neq 0$	$= 0$
Individual and interaction effects	$\neq 0$	$= 0$	$\neq 0$
Group and interaction effects	$= 0$	$\neq 0$	$\neq 0$
Individual, group, and interaction effects	$\neq 0$	$\neq 0$	$\neq 0$

3.4
The Use of Group Variables Other than Those Based on the Group Means

In the discussion of the basic model of contextual analysis, the group variable W is unspecified. An important group variable is subsequently defined as the within-group means for the individual-level explanatory variable X, and the analysis based on this variable is treated as the paradigm of contextual analysis. Nevertheless, there are group-level variables that do not have individual-level counterparts. These may at times become a vital part of the analysis. One likely prospect is an "integral variable" which is a variable representing the property of a group that cannot be derived from an individual-level variable.[5] Examples include such things as type of government and type of community.

Group properties such as type of government, gross national product, and welfare expenditures are defined as integral-type group variables because they cannot be derived from an individual-level property. In contrast, compositional-type variables are derived from an individual-level variable. The group mean is the prime example, but there are many others. Group size and the variance of X (homogeneity) are examples. Others include such things as population density and the Gini index of inequality.

An additional class of group variables not based on the means emerges with operations on several explanatory variables. These variables may be of the integral or the compositional type. Some examples are the factors which result from a factor analysis, or measures of group properties that result from multidimensional scaling procedures. Another important kind of group property might be based on the results of the description of relationships between individuals in a group, as in the sociometric analysis of communication networks.

The group means of the dependent variable is another type of explanatory variable. While this may sound curious at first, it does make sense in the contextual framework. Suppose, for example, that we are interested in predicting the probability of being poor, given one's race. It may be that the group variable racial composition would be a factor in this prediction. But it may also be of considerable interest to know how the level of poverty reflected in the group variable Y affects the various racial groups considered in the analysis. The model for this analysis is specified by the following equations:

$$y_{ik} = d_{0k} + d_{1k} x_{ik} + f_{ik} \tag{3.28}$$

$$d_{0k} = b_0 + b_2 \bar{x}_k + b_4 \bar{y}_k + u_k \tag{3.29}$$

$$d_{1k} = b_1 + b_3 \bar{x}_k + b_5 \bar{y}_k + v_k \tag{3.30}$$

The single-equation model is then

$$y_{ik} = b_0 + b_1 x_{ik} + b_2 \bar{x}_k + b_3 x_{ik} \bar{x}_k + b_4 \bar{y}_k + b_5 x_{ik} \bar{y}_k + e_{ik} \tag{3.31}$$

Generally speaking, in all the cases described above, other integral group variables would not replace the group-mean variable $\bar{X}$, but would be added to the basic model.

[5] The concept of "integral variables" is explained by Selvin and Hagstrom (1963).

3.5
The Problem of Multicollinearity in Measuring the Relative Importance of Individual, Group, and Interaction Effects

Two procedures for conceptualizing and estimating individual, group, and interaction effects are described above. Both procedures produce unstandardized regression coefficients that indicate the presence or absence of the corresponding multilevel effects. One problem in the interpretation of the coefficients arises because the ranges of the three variables will vary considerably. This is because the interaction variable is the product of the individual and group variables. In the single-equation procedure the range problem is solved by obtaining standardized regression coefficients.

A more serious problem arises because the explanatory variables are usually intercorrelated, making it difficult to untangle their unique contributions to the explanation of the dependent variable. This problem of multicollinearity is encountered in virtually all multivariate analyses. But because the individual-level variable X, the group variable $\bar{X}$, and the interaction variable $X\bar{X}$ are transformations of each other, it is especially serious in contextual analysis. This problem is addressed further in Chapters 4 and 13.

3.6
Summary

The basic model of contextual analysis for continuous variables is discussed in detail. This model specifies an individual-level dependent variable and a set of three explanatory variables consisting of an individual-level variable, a group variable based on group means, and an interaction variable based on the product of the individual and group variables.

Two different approaches to conceptualizing and estimating this model are developed. The first involves a two-step procedure beginning with a direct focus on the relationship between the individual-level variables within each group. In this approach, individual, group, and interaction effects are inferred from the magnitude of relationships found between the slopes and intercepts of the regression lines and the group means. The second approach involves the evaluation of a single multiple-regression equation containing terms for each of the three variables. The two approaches are introduced because they both have roots in the literature and because both are needed for subsequent discussion. It is observed that the two procedures will not always produce the same estimates of effects. This problem is taken up in Chapter 6 and in Appendix C.

From the basic three-variable model of contextual analysis, a simple typology of special cases is developed. The typology is based on the presence or absence of effects associated with the three variables. Estimates for the single-level equation involving only the individual-level variables are reported in the tables corresponding to each case. This dramatizes the point that analyses which ignore contextual variables can be misleading if the inference is that the results measure true individual-level effects.

The Measurement and Evaluation of Effects in a Contextual Analysis

The basic model of contextual analysis is specified in Chapter 3. Values of the unstandardized coefficients from this model are used to make inferences about the existence of individual, group, and interaction effects associated with them. The major consideration is whether the coefficients are different from zero. This chapter pursues the problem of evaluating the relative importance of the effects in explaining the dependent individual-level variable.

4.1 Analyzing the Variation Explained by Specified Variables

The data used to illustrate this discussion are presented in Table 3.1. Unstandardized coefficients for the individual, group, and interaction effects are presented in Table 4.1. One way to evaluate the relative importance of these effects is to compare the coefficients as they are. Given this approach, the group effect in the example appears most

Table 4.1
Coefficients Obtained from the Hypothetical Four-city Data

	CONSTANT	INDIVIDUAL EFFECT	GROUP EFFECT	INTERACTION EFFECT	R^2
Estimates from single-equation procedure	−13.72	3.28	8.03	0.34	0.764
Estimates after standardizing variables	—	0.26	0.45	0.22	0.764

Table 4.2
Sums of Squares for Stepwise Regressions for the Hypothetical Four-city Data

SOURCE	SUM OF SQUARES	PERCENT OF TOTAL VARIATION ACCOUNTED FOR
Individual	6241.0	61.3
Group, after individual	1521.0	14.9
Interaction, after individual and group	15.1	0.1
Residual	2406.6	23.6
Total	10,183.7	99.9

(a)

SOURCE	SUM OF SQUARES	PERCENT OF TOTAL VARIATION ACCOUNTED FOR
Group	6962.0	68.4
Individual, after group	800.0	7.8
Interaction, after group and individual	15.1	0.1
Residual	2406.6	23.6
Total	10,183.7	99.9

(b)

important. The individual effect is next, and the interaction effect appears negligible. However, as observed in Chapter 3, the ranges of the explanatory variables in the basic model are quite different because the interaction variable is made of the product of the individual and group variables. The problem might be remedied by standardizing the variables before making the estimates. Standardized estimates are presented in Table 4.1 along with the previous results. Standardizing the variables most notably affects the coefficient for the interaction effect. In this instance it becomes almost as large as the coefficient for the individual effect.

Unfortunately, standardizing the variables does not solve the more difficult problem of assessing the relative importance of the three effects.[1] Because the last two explanatory variables are constructed from the first variable, all will be highly correlated. In the four-city example the individual variable and the group variable have a correlation coefficient of 0.71, the individual and interaction variables 0.93, and the group and interaction variables 0.88. With or without standardization, unique portions of the variation in the dependent variable cannot be separated out and associated with each explanatory variable.

In such situations a common practice is to enter the unstandardized variables into the final equation one at a time and observe the successive reductions of the residual sum of squares. One possible order is the individual variable, the group variable, and the interaction variable. This involves regressing Y on X, then on X and $\bar{X}$, and finally on X, $\bar{X}$, and $X\bar{X}$. Using this order, the explained or regression sums of squares for the four-city data are shown in Table 4.2(a). For comparison, Table

[1] Issues involving the interpretation of relative importance of variables are discussed in Chapter 13.

4.2(b) shows the sums of squares resulting from first regressing Y on $\bar{X}$, then on $\bar{X}$ and X, and finally on all three variables.

When the individual-level variable is introduced first, it explains 6241.0/10183.8 = 61.3 percent of the total variation in Y. When the group variable is entered in the second position, it explains 14.9 percent of the variation. However, when the group variable is introduced first, it explains 68.4 percent of the total variation. After the individual and group variables have explained as much as they can, the interaction variable explains only 0.1 percent of the variation in Y. Yet it can be found that the interaction variable by itself explains 73.6 percent of the variation. Because the outcome depends on the order in which the variables are introduced, assessing the contributions of the three variables in a stepwise fashion does not solve the problem.

The stepwise procedure in which the individual-level variable is considered first also illustrates the point made in Section 3.1 about the problem of considering only an individual-level variable in the analysis of the dependent variable. A noncontext analysis of these data is tantamount to regressing Y on X for all sample observations without considering the effects of the group-mean variable $\bar{X}$. This might lead to the "questionable" inference that 61.3 percent of the variation in Y explained by X can be attributed solely to a property of individuals.

4.2

Analyzing the Variation Accounted for by Unspecified Variables and Interaction Variables

Typically, a substantial part of the total variation of the dependent variable will remain unexplained by a single individual-level variable and its derived values comprised of the group-mean variable and the interaction variable. Extending the basic model of contextual analysis to include additional sets of variables is discussed in Chapter 6. As in any analysis, identifying other relevant variables is primarily based on theory development and prior research. However, in contextual analysis it is assumed that some of the variation unexplained by the set of variables based on one individual-level variable is due to other individual-level variables, while some is due to other group and interaction variables. The ability to get an idea of where (on which level) most of the unexplained variation originates is diagnostically valuable in searching for other relevant variables.

An opportunity to explore the origins of unexplained variation is provided by the separate-equation procedure for estimating model coefficients. In contrast to the single-equation procedure, which minimizes only one error sum of squares e_{ik}, the separate-equation method minimizes three error sums. Analyzing these residuals, it is possible to partition the total unexplained variation into two parts. One is associated with unspecified individual-level variables, the other is associated with unspecified group and interaction variables together. This capability is illustrated using the four-city example.

One important set of residuals is obtained when the relationship between X and Y is analyzed within groups as in Eq. (3.3). The residuals are found from

$$f_{ik} = y_{ik} - d_{0k} - d_{1k}x_{ik} \tag{4.1}$$

For example, it is found that the residual sum of squares obtained from regressing Y on X in the first group equals 475.0. The way to measure the magnitude of all the residuals in all the groups is to add their sums of squares. For the example, we get

$$\sum_i \sum_k f_{ik}^2 = 475.0 + 450.0 + 475.0 + 825.0 = 2225.0 \tag{4.2}$$

The above residuals are all computed within the groups. The sums of squares therefore represent the unexplained part of the variation in the dependent variable due to individual-level variables.

A residual value of the dependent variable for the ith person in the kth group also results from the single-equation multiple-regression procedure. The sum of squares of the residuals from the multiple-regression analysis for the example is

$$\sum_i \sum_k e_{ik}^2 = 2406.6 \tag{4.3}$$

This sum of squares measures the unexplained part of the variation in Y due to *all* other (unspecified) individual, group, and interaction variables.

Two residual sums of squares are thereby identified. The larger one from Eq. (4.3) represents the unexplained variation due to all other unspecified variables. The smaller one from the several simple regressions in Eq. (4.2) represents the unexplained variation due to other (unspecified) individual-level variables. The difference

$$\sum_i \sum_k e_{ik}^2 - \sum_i \sum_k f_{ik}^2 = 2406.6 - 2225.0 = 181.6 \tag{4.4}$$

measures the unexplained variation in Y due to all other group and interaction variables.

Table 4.3 summarizes explained and unexplained variance for the substantive example. Column 2 shows the partitioning of the unexplained sums of squares just described. The explained variation is divided into two parts, one associated with the individual variable when it is entered first, and the remainder associated with the group and interaction variables. This column will change given different orderings of the variables. The column involving unexplained variation is not affected by the order in which specified variables are considered. The third column contains the sums of columns 1 and 2. This column provides a measure of how the total variation (explained and unexplained) is apportioned between the individual and the context (group and interaction) levels of observation.

An important connection between contextual analysis and the analysis-of-covariance approach described in Chapter 1 can be pointed out here. It is observed that the analysis-of-covariance model, with its nominal variable treatment of group

Table 4.3
Sum-of-squares Table for the Hypothetical Four-city Contextual Analysis

	Explained variation	Unexplained variation	Total explained and unexplained
Individual-level variable entered first	6241.0 80.2* 73.7 61.3	2225.0 92.4 26.3 21.8	8466.0 83.1
Group and interaction variables together	1536.1 19.8 89.4 15.1	181.6 7.5 10.6 1.8	1717.7 16.9
Total	7777.1 76.4	2406.6 24.6	10183.7 100.0

* Except for the marginal totals, percents are given first by column, then by row, and then corner.

effects, measures undifferentiated or composite group and interaction effects rather than effects associated with specific group and interaction variables. By contrast, contextual analysis associates group and interaction effects with specific variables. For a given case, the amount of variation explained by the specific group and interaction variables in the contextual model cannot be larger than the amount of undifferentiated group and interaction effects found with an analysis of covariance.

In the example the variation associated with group and interaction effects from the analysis of covariance (with the individual-level variable considered first) is 1717.7. Table 4.3 shows that this is the variation explained by the specified group and interaction variables in the contextual analysis *plus* the variation due to unspecified group and interaction variables.[2]

In the contextual-analysis approach, the amount of unspecified group and interaction effects is obtained from the separate equations involving within-group slopes and intercepts. Awareness of the relationship between the context model and the analysis-of-covariance model suggests the alternative of subtracting the group and interaction variation explained in a contextual analysis from the group and interaction variation explained in an analysis of covariance. This gives the amount of variation associated with group and interaction variables which are not specified in the context model.

The sums of squares table is summarized by calculating percentages down the columns, across the rows, and to the corner. Begin by looking down the columns. Observe that 6241.0/7777.1 = 80.2 percent of the explained variation in Y is accounted for by the individual-level variable. Of the total unexplained variation, 92.4 percent is accounted for at the individual level, and 83.1 percent of the total

[2] This is pointed out by Alwin (1976) and Firebaugh (1977).

variation in Y (explained and unexplained) is traceable to the individual level. This means that 19.8 percent of the explained variation is accounted for by unspecified group and interaction variables, 7.5 percent of the unexplained variation is accounted for by unspecified group and interaction variables, and 16.9 percent of the total variation is accounted for by group and interaction variables.

Looking across the table, the individual-level variable accounts for 6241.0/8466.0 = 73.7 percent of the variation found originating at the individual level. The group and interaction variables together account for 89.4 percent of the variation traceable to the context variable. Looking to the corner, the individual-level variable X accounts for 61.3 percent of the total variation in the dependent variable, and so forth.

From these results one would conclude that, of the three specified variables, the individual-level variable X contributes the most to the explained variation in Y (80.2 percent). However, at the individual level 26.3 percent of the total variation traceable to the individual level remains unexplained. In contrast, the group and individual variables together account for about 20 percent of the explained variation in Y, while only 7.5 percent of the total variation originating at the group and interaction levels remains unexplained by the specified group and interaction variables. This suggests that a search for additional variables should concentrate on individual-level variables. The problem remains, however, that these results were strongly influenced by the decision to consider the individual variable first in calculating the sums of squares.

4.3

A Centering Procedure to Evaluate the Unique Contribution of the Individual, Group, and Interaction Variables

Because the three explanatory variables in the basic model are transformations of each other, they will generally be highly correlated. The procedures which are proposed are designed to remove these correlations. The rationale for these procedures is comparable to the rationale for having equal numbers of subjects in experimental groups. That is, the purpose is to make independent variables orthogonal. These procedures are illustrated using the four-city data from Table 3.1. Computer operations are described in Appendix A, Illustrations 8 and 9.

Because the essential information for the individual, group, and interaction effects is contained in the group intercepts and slopes of the regression lines, the values of the intercepts and slopes should be preserved. Moving the regression lines relating Y and X in each group so that the mean points on the lines are located above each other on the Y axis achieves what is desired without interfering with the values of the intercepts and slopes. Specifically, each point (x_{ik}, y_{ik}) is moved along the regression line of the group in which the point belongs. Figure 4.1 shows how the

Figure 4.1
Centering a Regression Line to Make the Mean of the Explanatory Variable Equal Zero for the kth Group while Preserving the Same Slope and Intercept

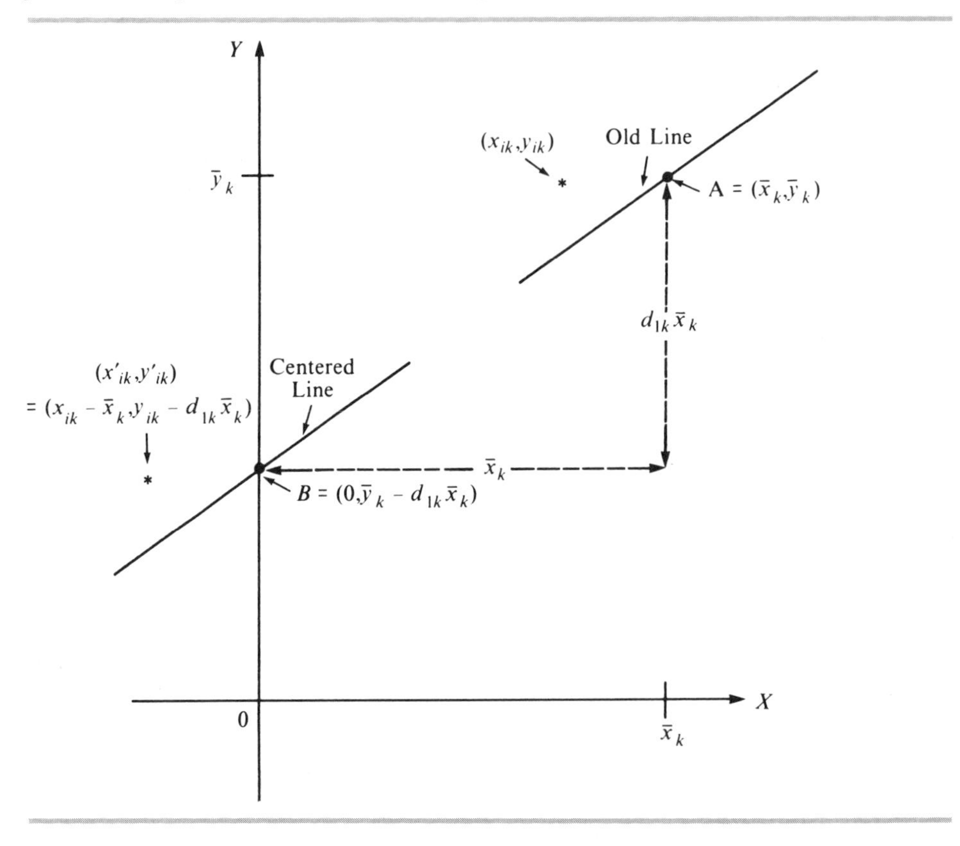

regression line is moved for the first city in Table 3.1. Consider the line containing the mean point marked A with coordinates $(\bar{x}_k,\bar{y}_k)$. The values of these coordinates in the example are 3 and 21.5. The objective is to move this line so that the intercept and slope are not disturbed and so that point A falls on the Y axis at the point marked B. This is done by moving point A to the left a distance of $\bar{x}_k$. The new X coordinate is obtained by subtracting the value $\bar{x}_k$ from the old X coordinate. Thus the new X coordinate becomes $\bar{x}_k - \bar{x}_k = 0$. Similarly, the point A is moved down a distance of $d_{1k}\bar{x}_k$ since the slope of the line equals d_{1k}. To obtain the new Y coordinate, subtract the value $d_{1k}\bar{x}_k$ from the old Y coordinate. The Y value for the point B becomes $\bar{y}_k - d_{1k}\bar{x}_k$.

The last step is to move all the observations in the scatter plot around the old line in the same way and to denote the new coordinates for each point with a prime. The new coordinates then relate to the old coordinates according to the equations

$$x'_{ik} = x_{ik} - \bar{x}_k, \qquad x_{ik} = x'_{ik} + \bar{x}_k \tag{4.5}$$

$$y'_{ik} = y_{ik} - d_{1k}\bar{x}_k, \qquad y_{ik} = y'_{ik} + d_{1k}\bar{x}_k \tag{4.6}$$

Computer procedures are described in Appendix A, Illustration 8. Consider the first observation in city 1 in Table 3.1. The value of x_{11} equals 1, and y_{11} equals 19. The mean of X in that city equals 3, and the slope equals 2.5. These new coordinates are

$$x'_{11} = 1 - 3 = -2 \tag{4.7}$$

$$y'_{11} = 19 - 2.5(3) = 11.5 \tag{4.8}$$

The sixteen transformed observations for the dependent and three explanatory variables are presented in Table 4.4.

Table 4.4
Transformed Variables after Centering Operations for the Hypothetical Four-city Data

CITY NUMBER k	DEPENDENT VARIABLE y'	INDIVIDUAL VARIABLE x'	GROUP VARIABLE $\bar{x}'_k$	INTERACTION VARIABLE $x\bar{x}'_k$
1	11.5	−2	−2	4
	26.5	0	−2	0
	−3.5	0	−2	0
	21.5	2	−2	−4
2	14	−2	0	0
	41	0	0	0
	11	0	0	0
	38	2	0	0
3	25	−2	0	0
	43	0	0	0
	13	0	0	0
	41	2	0	0
4	15.5	−2	2	−4
	45.5	0	2	0
	5.5	0	2	0
	45.5	2	2	4

The uncentered and centered four-city regression lines are presented in Figure 4.2. The original lines are shown in Figure 4.2(a). The intercepts and slopes in Figure 4.2(b) are the same after centering as before. However, all the new means of the explanatory variables now equal zero and the dependent variable has new values.

Identical in both graphs, the interaction effect shows up in Figure 4.2 as unequal slopes, the group effect shows up in the unequal intercepts, and the individual effect shows up in the nonzero slopes. As a result of this data adjustment, the regression lines are now *centered* on the Y axis.

Now we need to express the new lines mathematically. The relationship between the variables X and Y in the kth group is specified in Chapter 3 [Eq. (3.3)]. To repeat,

Figure 4.2
Comparison of Uncentered and Centered Regression Lines, Showing the Within-group Slopes and Intercepts Preserved

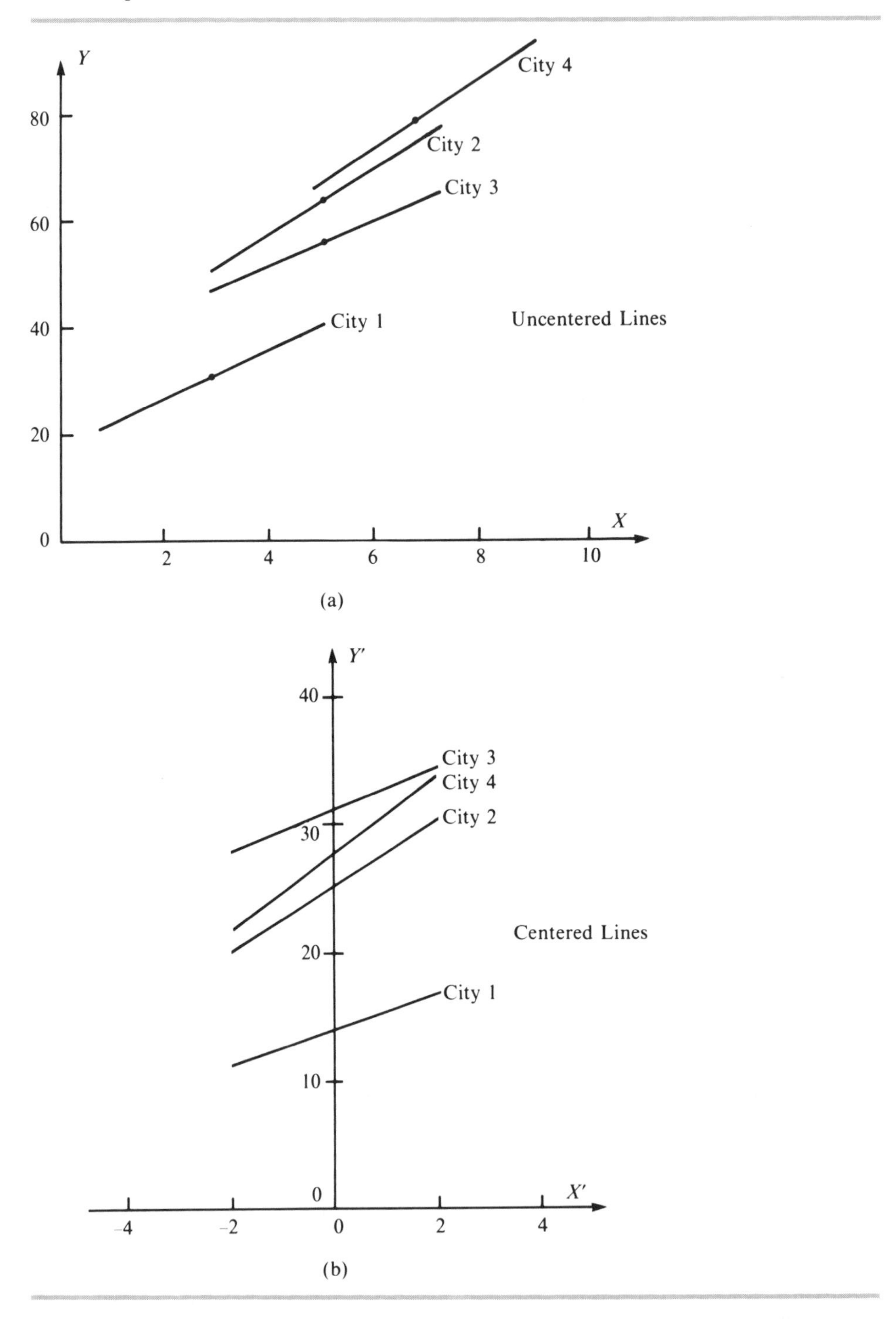

this equation is

$$y_{ik} = d_{0k} + d_{1k}x_{ik} + f_{ik} \tag{4.9}$$

Substituting y_{ik} of Eq. (4.6) into Eq. (4.9), gives

$$y'_{ik} + d_{1k}\bar{x}_k = d_{0k} + d_{1k}x_{ik} + f_{ik} \tag{4.10}$$

Keeping y'_{ik} on the left-hand side of Eq. (4.10), the relationship between the new variable Y' and the centered X variable is expressed in the equation

$$y'_{ik} = d_{0k} + d_{1k}(x_{ik} - \bar{x}_k) + f_{ik} \tag{4.11}$$

where $x_{ik} - \bar{x}_k = x'_{ik}$.

It is possible to obtain centered estimates from the separate-equation procedure involving the intercepts and the slopes in a similar fashion. The separate-equation procedure specifies that the intercepts d_{0k} and the slopes d_{1k} are functions of the group means. Centering involves expressing these quantities as functions of the group means minus the overall mean. Expressed mathematically this is

$$d_{0k} = a_0 + a_2(\bar{x}_k - \bar{x}) + u_k \tag{4.12}$$

$$d_{1k} = a_1 + a_3(\bar{x}_k - \bar{x}) + v_k \tag{4.13}$$

Computer procedures are described in Appendix A, Illustration 9. When the group sizes vary considerably, estimates for these equations should be obtained through weighted least-square procedures. Details of this weighting procedure are contained in Appendix B.

To indicate that this is now the centered model, the b's are changed to a's. To show the relationships between the a's and the b's, substitute d_{0k} and d_{1k} from Eqs. (4.12) and (4.13) into the uncentered separate equations, Eqs. (3.6) and (3.7). Thus,

$$\begin{aligned} b_0 &= a_0 - a_2\bar{x} \\ b_1 &= a_1 - a_3\bar{x} \\ b_2 &= a_2 \\ b_3 &= a_3 \end{aligned} \tag{4.14}$$

Notice that the intercepts in the uncentered separate equations (b_0 and b_1) change with the centering procedure, but the slopes (b_2 and b_3) remain the same.

The centered-model single equation is obtained in the same manner as the uncentered model. That is, substituting the d's from Eqs. (4.12) and (4.13) into Eq. (4.11) gives

$$y'_{ik} = a_0 + a_1(x_{ik} - \bar{x}_k) + a_2(\bar{x}_k - \bar{x}) + a_3(x_{ik} - \bar{x}_k)(\bar{x}_k - \bar{x}) + e'_{ik} \tag{4.15}$$

where the residuals $e'_{ik} = f_{ik} + u_k + v_k(x_{ik} - \bar{x}_k)$.

Numerical estimates of effects in the four-city data obtained from the separate- and single-equation estimation procedures for the centered model of contextual analysis are contained in Table 4.5. For comparison, the separate- and single-equation estimates from the uncentered model considered previously are also presented. These results show that the two methods for estimating effects in the uncentered model can differ substantially. Note that the corresponding estimates of individual-level effects from the uncentered model are of opposite signs. This problem is addressed in Chapter 6 under the discussion of nonlinear context effects. In contrast, estimates from the two procedures in the centered model yield identical results. This reconciliation of the separate- and single-equation approaches in contextual analysis is an important consequence of the centering procedure and will generally occur when the within-group variances of X do not vary substantially. This point is elaborated in Appendix D.

Table 4.5
Coefficients Obtained from the Uncentered and Centered Separate- and Single-equation Procedures

	CONSTANT	INDIVIDUAL EFFECT	GROUP EFFECT	INTERACTION EFFECT
Estimates from the uncentered model	b_0	b_1	b_2	b_3
Separate-equation method	7.13	−1.25	3.50	1.25
Single-equation method	−13.72	3.28	8.03	0.34
Estimates from the centered model	a_0	a_1	a_2	a_3
Separate-equation method	24.63	5.00	3.50	1.25
Single-equation method	24.63	5.00	3.50	1.25

4.3.1. Partitioning Explained Variation after Centering

Another important consequence of the centering procedure is that the correlations between the transformed explanatory variables are equal to zero as desired. This too will generally occur with the centering procedure when the variances of X within each group do not vary substantially. See Appendix D for a discussion of this issue.

Removing the correlations between the explanatory variables while preserving the within-group intercepts and slopes makes it possible to partition the regression sums of squares (representing the total explained variation) into three unique parts, one for each variable. Thus a sum-of-squares table will have a column for the total amount of explained variation and for the amounts associated with each variable.

With the explanatory variables uncorrelated, the partial regression coefficients in the single equation are identical to the zero-order coefficients relating each variable to the dependent variable separately. For this one-variable-at-a-time procedure we have the predicted values of Y (denoted by a caret).

For the individual-level variable,

$$\hat{y}'_{ik} = a_0 + a_1(x_{ik} - \bar{x}_k) \tag{4.16}$$

For the group-level variable,

$$\hat{y}'_{ik} = a_0 + a_2(\bar{x}_k - \bar{x}) \tag{4.17}$$

For the interaction variable,

$$\hat{y}'_{ik} = a_0 + a_3(x_{ik} - \bar{x}_k)(\bar{x}_k - \bar{x}) \tag{4.18}$$

Formulas for the sums associated with these equations are given in Table 4.6, making it possible to obtain their values through the separate-equation procedure. Numerical values of these sums of squares for the four-city example are presented in Table 4.7.

The relative contributions of the three explantory variables can be interpreted in terms of proportions of the total explained variation. Calculating down the first column of Table 4.7, the individual-level variable accounts for $800/1292 = 61.9$ percent of the total explained variation in Y. The group variable accounts for 30.3 percent of the explained variation. The interaction variable accounts for 7.7 percent of the variation.

The effort to deal with multicollinearity in this particular contextual analysis shows that group and interaction effects are more important than indicated with an arbitrary stepwise procedure in which the individual-level variable is considered first. On the other hand, group and interaction effects are considerably less important than indicated if either one was arbitrarily considered first in a stepwise procedure.

Table 4.6
Sum-of-squares Table with Formulas for the Variation after Centering

	Explained by specified variables	Unexplained by specified variables	Total
Individual level	$a_1^2\sum\sum(x_{ik} - \bar{x}_k)^2$	$\sum\sum f_{ik}^2$	$\sum\sum(y'_{ik} - \bar{y}')^2 - \sum(d_{1k} - a_1)^2\sum(x_{ik} - \bar{x}_k)^2$
Group level	$a_2^2\sum n_k(\bar{x}_k - \bar{x})^2$	$\sum n_k u_k^2$	$\sum n_k(d_{0k} - a_0)^2$
Interaction level	$a_3^2\sum\sum(x_{ik} - \bar{x}_k)^2(\bar{x}_k - \bar{x})^2$	$\sum v_k^2\sum(x_{ik} - \bar{x}_k)^2$	$\sum(d_{1k} - a_1)^2\sum(x_{ik} - \bar{x}_k)^2$
Total	$\sum\sum(\hat{y}'_{ik} - \bar{y}')^2$	$\sum\sum e_{ik}^2$	$\sum\sum(y'_{ik} - \bar{y}')^2$

Table 4.7
Sum-of-squares Table with the Centered Model for the Hypothetical Four-city Data

	Explained variation	Unexplained variation	Total explained and unexplained
Individual level	800.0 61.9* 26.4 21.1	2225.0 89.3 73.6 58.8	3025.0 79.9
Group level	392.0 30.3 61.0 10.4	250.8 10.1 39.0 6.6	642.8 17.0
Interaction level	100.0 7.7 86.2 2.6	16.0 0.6 13.8 0.4	116.0 3.1
Total	1292.0 34.1	2491.8 65.9	3783.8 100.0

* Except for the marginal totals, percents are given first by column, then by row, and then corner.

4.3.2. Partitioning Unexplained Variation after Centering

Another major consequence of applying the centering procedure is that the variation unexplained by the three variables can be divided into three unique parts and attributed to other unspecified individual variables, group variables, and interaction variables. Centering thereby serves an important diagnostic function in the search for other explanatory variables.

Three sets of residuals follow from the centered model. One set of residuals is obtained when the relationship between X and Y is examined within the groups. These residuals are designated f_{ik}. The relationship involving these residuals is expressed in Eq. (4.11). The residual term f_{ik} is found from

$$f_{ik} = y'_{ik} - d_{0k} - d_{1k}(x_{ik} - \bar{x}_k) \qquad (4.19)$$

Since y'_{ik} equals $y_{ik} - d_{ik}\bar{x}_k$ [from Eq. (4.6)], we can see that this is the same f_{ik} produced by the uncentered model. The centering procedure does not disturb the within-group slopes and intercepts nor the position of the observations in relation to their line.

To measure the magnitude of all the f_{ik} residuals in all the groups, we computed their sum of squares in Eq. (4.2). For the four-city example, this quantity is 2225.0. This value appears in the second column of Table 4.7. Residuals are computed within all the groups, and the sum of squares therefore represents the unexplained part of the variation in the dependent variable due to unspecified individual-level variables.

The residual value of the dependent variable for the ith person in the kth group resulting from the single-equation multiple-regression analysis appears in Eq. (4.15) and is obtained from

$$e'_{ik} = y'_{ik} - a_0 - a_1(x_{ik} - \bar{x}_k) - a_2(\bar{x}_k - \bar{x}) - a_3(x_{ik} - \bar{x}_k)(\bar{x}_k - \bar{x}) \tag{4.20}$$

The sum of squares of the residuals from the single-equation regression analysis of the four-city data is

$$\sum_i \sum_k e'_{ik} = 2491.8 \tag{4.21}$$

This residual sum of squares measures the unexplained part of the variation in Y due to all unspecified individual, group, and interaction variables. The numerical value of this quantity appears at the bottom of the second column in Table 4.7.

There now are two measures of unexplained variation. The larger one from the three-variable multiple-regression analysis gives the unexplained variation due to all unspecified variables. The smaller one from the several simple regressions gives the unexplained variation due to all unspecified individual-level variables. The difference

$$\sum_i \sum_k e'^2_{ik} - \sum_i \sum_k f^2_{ik} = 2491.8 - 2225.0 = 266.8 \tag{4.22}$$

measures the unexplained variation in Y due to all unspecified group and interaction variables.

These observations separate the purely individual-level sources of unexplained variation from group and interaction sources, but they do not separate the group and interaction sources. Although not possible in the uncentered model, it is possible in the centered model. Begin with the separate equations for the centered model. The relationships involving the within-group intercepts and slopes for the centered model are specified in Eqs. (4.12) and (4.13). The residuals u_k and v_k in these equations are the same as in the uncentered model. However, the sums of squares associated with them differ because of the way they appear in the single equation for the centered model. From Eq. (4.15) we get

$$e'_{ik} = f_{ik} + u_k + v_k(x_{ik} - \bar{x}_k) \tag{4.23}$$

The residuals shown in Eq. (4.23) consist of three unique components. The first component represents the effect on Y of all other individual-level variables, the second represents the effect of all other group-level variables, and the third represents the effect of all other interaction-level variables. After centering these components are unique, because they are uncorrelated with each other. Again, this will generally be the case when the within-group variances of X do not vary substantially. These conditions are discussed in Appendix D.

The contributions of the various levels of unexplained variation are summarized by squaring Eq. (4.23) and summing across all individuals and groups. Because the three terms are uncorrelated, this gives

$$\sum_i \sum_k e_{ik}'^2 = \sum_i \sum_k f_{ik}^2 + \sum_k n_k u_k^2 + \sum_k v_k^2 \sum_i (x_{ik} - \bar{x}_k)^2 \qquad (4.24)$$

Formulas for each of the three parts of the unexplained variation are given in the second column of Table 4.6. From Table 4.7 it can be observed that the total amount of unexplained variation is equal to $2225.0 + 250.8 + 16.0 = 2491.8$.

Interpreting the total residual sum of squares has already been done for the first component, $\sum\sum f_{ik}^2 = 2225.0$. This is the part of the variation in Y due to other individual-level variables. The next component, $\sum n_k u_k^2 = 250.8$, is the unexplained variation in Y on the group level due to other group variables. The final term, $\sum v_k^2 \sum (x_{ik} - \bar{x}_k)^2 = 16.0$, is the unexplained variation in Y due to other interaction-level variables.

One way to summarize the unexplained variation in the expanded sum-of-squares table for the centered model is by looking down the columns. Down the second column it can be observed that other individual-level variables account for 89.3 percent of the total unexplained variation in the dependent variable. Other group variables account for 10.1 percent of the unexplained variation. Other interaction variables account for 0.6 percent of the unexplained variation. The analysis of the unexplained variation in this example suggests that most of the variance left unexplained by the basic contextual model is traceable to the individual level, and that is where attention should be directed for the inclusion of other variables into the analysis.

4.3.3. Partitioning Total Variation (Explained and Unexplained) by Individual, Group, and Interaction Sources

The third column of Tables 4.6 and 4.7 is obtained by adding columns 1 and 2. This produces a decomposition of the variation of Y' into three parts, each corresponding to one of the levels of observation and each consisting of explained *and* unexplained variance. From this angle all the variation in the dependent variable is considered to be accounted for, even if all explanatory variables have not been explicitly measured. The question is where this variation originates in terms of levels.

The quantities called for in the third column of Table 4.7 can also be obtained using the separate-equation procedure and the formulas specified there. These formulas and the partitioning of sums of squares are elaborated in Appendix D.

Numerical results for the substantive example in this decomposition of variation are presented in the third column of Table 4.7. As before, these results can be summarized by computing the proportions looking down the column to the cell marked total. Thus the individual level accounts for $3025.0/3783.8 = 79.9$ percent of the total variation, the group level accounts for 17.0 percent of the total variance, and the interaction level accounts for 3.1 percent of the total variation.

Having filled in the squares of the expanded analysis, other properties of this particular contextual structure can be highlighted by reporting proportions computed across the rows of Table 4.7. Begin with the row marked total, where the explained sum of squares accounts for 1292.0/3783.8 = 34.1 percent of the total variation. This can be seen to be a simple breakdown of the total sums of squares according to a regular three-variable regression analysis. At the individual level, 800.0/3025.0 = 26.4 percent has been explained and 73.6 percent is unexplained. At the group level, 61.0 percent has been explained and 39.0 percent is unexplained. Finally, at the interaction level, 86.2 percent has been explained and 13.8 percent is unexplained.

To examine the results of the analysis from yet another angle, one can also divide the cells in Table 4.7 by the corner. For the cell corresponding to the individual-level and explained variation, 800.0/3783.8 = 21.1 percent of the total variation. For the cells under explained variation this procedure will produce the familiar R^2 values for the simple regression between Y' and each of the three explanatory variables. Because the explanatory variables are uncorrelated, the square roots of these R^2 values will be identical to the standardized regression coefficients (or beta weights) in the single equation. These are commonly used to eliminate the dependence of the coefficients on the scales of the explanatory variables. Standardizing the data to obtain the corresponding standardized coefficients, we get

$$z'_y = 0.46z_{\text{ind}} + 0.33z_{\text{grp}} + 0.16z_{\text{interact}} + z_e \tag{4.25}$$

From this analysis of variation of the three specified variables, one might conclude that the individual-level variable (tolerance) contributes most to understanding the variable Y', followed by the group variable, and then the interaction variable. However, compare these results to the previous results obtained from the stepwise analysis of variation in which the individual-level variable is considered first (Table 4.2). Notice that the group variable is more important in explaining deviance than previously indicated. At the same time, the results of the centering procedure clearly indicate that the group or interaction variables are not nearly as important as they might appear in a stepwise analysis in which they were considered before the individual-level variable. The sum-of-squares table also indicates that unspecified individual-level variables account for almost 90 percent of the variation left unexplained by the basic model. This suggests that the search for additional explanation in the example should concentrate on other individual-level variables.

4.4
Summary

Evaluating effects in the basic model of contextual analysis due to multicollinearity in the explanatory variables is explored at length. To partition the explained variation in the dependent variable between the three explanatory variables while ignoring the correlations between the variables, one must decide on a stepwise order in which to introduce them into the final three-variable regression. This decision will often produce the same ordering in terms of importance as the order in which they were introduced. To remedy the problem, a centering procedure is proposed which removes the correlations between the explanatory variables before analyzing their contributions to the dependent variable. It is then possible to divide the explained variation into three unique parts, one associated with each variable.

Also proposed is a procedure which divides the variation left *unexplained* by the single set of explanatory variables. This unique diagnostic aid is made possible by the separate-equation approach. In the uncentered model the unexplained variation can be divided into two parts, one traceable to unspecified individual-level variables and one to unspecified group and interaction variables together. With an application of the centering procedure, the unexplained variation can be broken up into three unique parts, one associated with unspecified individual-level variables, one associated with unspecified group variables, and one associated with unspecified interaction variables. The results of this partitioning of the unexplained variation direct the analyst to the most promising level at which to look for additional variables.

Contextual Analysis with Categorical Individual-level Data

Frequently an analysis problem involves categorical rather than continuous individual-level variables. For example, the problem may require a contextual analysis of the relationship between religious affiliation and choices between political candidates. In this chapter it is first demonstrated how a dichotomous-variable situation can be treated as a special case of the basic model of contextual analysis for continuous data described in Chapters 3 and 4. This is followed by a discussion of the case involving multiple-category individual-level variables. Related computer procedures are contained in Appendix A.

5.1 *Contextual Analysis with 2 × 2 Contingency Tables*

The example used to illustrate the procedures for dichotomous data is based on data from a study of normative constraints on deviant behavior in college contexts (cf. Bowers, 1968). The dichotomous dependent variable has two categories, getting drunk and not getting drunk. The individual-level explanatory variable consists of disapproving of getting drunk and not disapproving of getting drunk. The groups consist of five groups of colleges (in subsequent discussion, the term "colleges" refers to groups of colleges). The theory being explored is that individuals attending colleges where getting drunk is generally disapproved are themselves less likely to get drunk, regardless of whether they personally approve or disapprove of getting drunk.

5.1.1. Notation for 2 × 2 Tables

Each dichotomous variable in the analysis is treated as a dummy variable, with the categories assigned values 0 and 1. Table 5.1 shows this assignment of values to

Table 5.1
X and Y as Dummy Variables in the kth Table Together with the Notation for Frequencies, Proportions, and Column Conditional Proportions

Frequencies

		X: 0	X: 1	Sum
Y	1	n_{11k}	n_{12k}	$n_{1\cdot k}$
	0	n_{21k}	n_{22k}	$n_{2\cdot k}$
	Sum	$n_{\cdot 1k}$	$n_{\cdot 2k}$	n_k

Proportions

		X: 0	X: 1	Sum
Y	1	p_{11k}	p_{12k}	$p_{1\cdot k}$
	0	p_{21k}	p_{22k}	$p_{2\cdot k}$
	Sum	$p_{\cdot 1k}$	$p_{\cdot 2k}$	1.00

Column conditional proportions

		X: 0	X: 1
Y	1	m_{11k}	m_{12k}
	0	m_{21k}	m_{22k}
	Sum	1.00	1.00

individuals in the various categories of X and Y, together with the notation used for the number of observations and proportions in the cells in the kth table.

The particular layout of tables in Table 5.1 is designed to be consistent with the plotting of relationships in regression analysis. Thus X is taken to be the horizontal variable with 0 assigned to the category on the left, and Y is taken as the vertical variable with 0 assigned to the lower category. The proportions are obtained by dividing the frequencies by n_k, the total number of observations. The conditional proportions are obtained by dividing each frequency by the total number of observations in the column in which the cell is located. This notation is illustrated in Table 5.2 with observations for the first college in the substantive example. Table 5.3 contains data for each of the five colleges. The contingency tables in Table 5.3 contain the cell and marginal frequencies, the conditional proportions (in parentheses), and the marginal proportions.

Sum the frequencies n_{11k}, $n_{1\cdot k}$, $n_{\cdot 1k}$, etc., in Table 5.1 over the K groups to obtain the population or the sum table of frequencies. From this can be obtained, as above, the sum table of proportions and conditional proportions. The sum table is distinguished notationally from the individual-group tables by the absence of the subscript k as in Table 5.4. This table also contains the corresponding numerical entries for the five-college example.

Table 5.2
Data for College 1 in Table Form

	GETTING DRUNK Y		DISAPPROVAL OF GETTING DRUNK X		
Frequencies			X		
			0	1	Sum
	Y	1	561	15	576
		0	206	108	314
		Sum	767	123	890
Proportions			X		
			0	1	Sum
	Y	1	0.630	0.017	0.647
		0	0.231	0.121	0.353
		Sum	0.862	0.138	1.000
Column conditional proportions			X		
			0	1	
	Y	1	0.731	0.122	
		0	0.269	0.878	
		Sum	1.000	1.000	

5.1.2. Applying the Basic Model of Contextual Analysis to Dichotomous Variables

The formal model of contextual analysis presented in Chapter 3 specifies that the variables Y and X are related according to the equation

$$y_{ik} = d_{0k} + d_{1k}x_{ik} + f_{ik} \tag{5.1}$$

It further specifies that the intercepts and slopes are functions of the contextual variable $\bar{X}$ according to the equations

$$d_{0k} = b_0 + b_2\bar{x}_k + u_k \tag{5.2}$$

$$d_{1k} = b_1 + b_3\bar{x}_k + v_k \tag{5.3}$$

When d_{0k} and d_{1k} are substituted into Eq. (5.1), this produces the single equation

$$y_{ik} = b_0 + b_1x_{ik} + b_2\bar{x}_k + b_3(x_{ik}\bar{x}_k) + (f_{ik} + u_k + v_kx_{ik}) \tag{5.4}$$

Table 5.3

Data for Each of the Five Colleges in the Study of Social Norms and Deviance with Frequencies, Column Conditional Proportions, and Marginal Proportions

GETTING DRUNK Y — DISAPPROVAL OF GETTING DRUNK X

College 1

Y	X = 0	X = 1	
1	561 (0.731)	15 (0.122)	576 0.647
0	206 (0.269)	108 (0.878)	314 0.353
	767 0.862	123 0.138	890

College 2

Y	X = 0	X = 1	
1	476 (0.683)	42 (0.111)	518 0.481
0	221 (0.317)	338 (0.889)	559 0.519
	697 0.647	0.380 0.353	1077

College 3

Y	X = 0	X = 1	
1	525 (0.558)	80 (0.080)	605 0.312
0	416 (0.442)	920 (0.920)	1336 0.688
	941 0.485	1000 0.515	1941

College 4

Y	X = 0	X = 1	
1	125 (0.425)	27 (0.040)	152 0.158
0	169 (0.575)	644 (0.960)	813 0.842
	294 0.305	671 0.695	965

College 5

Y	X = 0	X = 1	
1	18 (0.360)	4 (0.010)	22 0.051
0	32 (0.640)	379 (0.990)	411 0.949
	50 0.115	383 0.885	433

Source: Adapted from Bowers, 1968, p. 397.

Table 5.4
Dummy Variable Notation for the Population (Sum Table) with Numerical Values for the Five-college Example

GETTING DRUNK Y — DISAPPROVAL OF GETTING DRUNK X

Frequencies

Y	X = 0	X = 1			Y	X = 0	X = 1	
1	n_{11}	n_{12}	$n_{1\cdot}$		1	1705	168	1873
0	n_{21}	n_{22}	$n_{2\cdot}$		0	1044	2389	3433
	$n_{\cdot 1}$	$n_{\cdot 2}$	n			2749	2557	5306

Proportions

Y	X = 0	X = 1			Y	X = 0	X = 1	
1	p_{11}	p_{12}	$p_{1\cdot}$		1	0.321	0.032	0.353
0	p_{21}	p_{22}	$p_{2\cdot}$		0	0.197	0.450	0.647
	$p_{\cdot 1}$	$p_{\cdot 2}$	1.00			0.518	0.482	1.000

Column condition proportions

Y	X = 0	X = 1		Y	X = 0	X = 1
1	m_{11}	m_{12}		1	0.620	0.066
0	m_{21}	m_{22}		0	0.380	0.934
	1.00	1.00			1.000	1.000

The coefficient b_1 provides a measure of the individual effect, b_2 provides a measure of the group effect, and b_3 provides a measure of the interaction effect. The case involving dichotomous variables can be developed as a special case of this general model in the following way.

When $n._{1k}$ individuals are assigned the value of 0 on X and $n._{2k}$ individuals are assigned the value of 1 on X, then the mean of X equals the proportion of individuals assigned the value 1, that is,

$$\bar{x}_k = \frac{n._{1k}(0) + n._{2k}(1)}{n_k} = p._{2k} \tag{5.5}$$

For college 1 in Table 5.3 this value is 0.138. Thus 13.8 percent of the students in college 1 disapprove of getting drunk. Similarly, the mean of Y becomes

$$\bar{y}_k = p_{1 \cdot k} \tag{5.6}$$

which is the proportion of individuals in the first row of the kth table. For college 1 this value is 0.647. Thus 64.7 percent of the students in college 1 did get drunk.

Since any estimated regression line goes through the point with coordinates $(\bar{x}, \bar{y})$, we get from Eq. (5.1) the following equation relating the two means, intercept and slope,

$$\bar{y}_k = d_{0k} + d_{1k}\bar{x}_k \tag{5.7}$$

Using the expressions for the two means already derived, we get the following equation for the relationship between the marginal proportions, intercept and slope:

$$p_{1 \cdot k} = d_{0k} + d_{1k} p_{\cdot 2k} \tag{5.8}$$

Finally, the intercept and slope for two dummy variables, as they have been defined here, can be expressed in terms of the two conditional column proportions. Among individuals assigned the value of 0 on X, the proportion m_{11k} has the value 1 on Y. Similarly, among individuals assigned the value 1 on X, the proportion m_{12k} has value 1 on Y. That is, the two conditional column proportions become

$$m_{11k} = \frac{n_{11k}}{n_{\cdot 1k}}, \qquad m_{12k} = \frac{n_{12k}}{n_{\cdot 2k}} \tag{5.9}$$

as shown in Table 5.1. The observed values in college 1 are 0.731 and 0.122. This indicates that 73.1 percent of those students who did not disapprove of getting drunk got drunk while 12.2 percent of those who disapproved got drunk.

From the way the frequencies add up in each of the tables, the row total for the first row equals the sum of the two corresponding cell frequencies, that is,

$$n_{1 \cdot k} = n_{11k} + n_{12k} \tag{5.10}$$

For example, in college 1, $576 = 561 + 15$. Substituting n_{11k} and n_{12k} from Eq. (5.9) into Eq. (5.10),

$$n_{1 \cdot k} = m_{11k} n_{\cdot 1k} + m_{12k} n_{\cdot 2k} \tag{5.11}$$

Dividing each side of Eq. (5.11) by n_k and taking advantage of the fact that $n_{1 \cdot k}/n_k = p_{1 \cdot k}$, $n_{\cdot 1k}/n_k = p_{\cdot 1k} = 1 - p_{\cdot 2k}$, and $n_{\cdot 2k}/n_k = p_{\cdot 2k}$,

$$p_{1 \cdot k} = m_{11k}(1 - p_{\cdot 2k}) + m_{12k} p_{\cdot 2k} \tag{5.12}$$

Finally, by carrying out the multiplication in Eq. (5.12), factoring, and rearranging the terms,

$$p_{1 \cdot k} = m_{11k} + (m_{12k} - m_{11k}) p_{\cdot 2k} \tag{5.13}$$

Comparing Eqs. (5.8) and (5.13), note that the intercept d_{0k} equals the conditional proportion in the first column, m_{11k}. Similarly, the slope d_{1k} equals the difference between the two conditional proportions. Thus,

$$\begin{array}{cc} \text{intercept} & \text{slope} \\ m_{11k} = d_{0k}, & m_{12k} - m_{11k} = d_{1k} \end{array} \tag{5.14}$$

Table 5.3 indicates that these values for college 1 are 0.731 and (0.122 − 0.731) = −0.609. Substituting the d's from Eq. (5.14) into Eq. (5.1), we obtain the 2 × 2 table equivalent of the within-group equations relating Y and X. Thus,

$$y_{ik} = m_{11k} + (m_{12k} - m_{11k})x_{ik} + f_{ik} \tag{5.15}$$

By the same reasoning, the population intercept and slope (in the sum table) can be found from

$$m_{11} = d_{0p}, \qquad m_{12} - m_{11} = d_{1p} \tag{5.16}$$

where the subscript p indicates that the coefficient is for the whole population. In the example the population consists of all five colleges. Table 5.4 shows that these values are 0.620 and −0.554, respectively.

Another important equivalence between the language used in Chapter 3 and the language of 2 × 2 contingency tables is pointed out here for later reference. That is, the variances of X and Y and the product of the marginal proportions in the tables are equivalent. For the variance of Y simply calculate the product of the marginal proportions associated with the dependent variable. Representing the variance of Y for the kth group by s_{yk}^2,

$$s_{yk}^2 = p_{1 \cdot k} p_{2 \cdot k} \tag{5.17}$$

For the variance of X,

$$s_{xk}^2 = p_{\cdot 2k} p_{\cdot 1k} \tag{5.18}$$

In college 1 these values are 0.228 and 0.119. For the population variances

$$s_y^2 = p_{1 \cdot} p_{2 \cdot}, \qquad s_x^2 = p_{\cdot 2} p_{\cdot 1} \tag{5.19}$$

these values are 0.228 and 0.249.

Thus the within-group and the population intercepts, slopes, means, and variances are simple to compute when the data consist of a set of 2 × 2 contingency tables. This makes it possible to find the coefficients in the contextual model, using the separate-equation procedure directly from the marginal and conditional proportions. To be specific, from Eqs. (5.2) and (5.3) the formal model has the intercepts d_{0k} and the slopes d_{1k} as functions of the group variable $\bar{X}$. Here we want to express

Table 5.5
Observed Values of m_{11k}, $m_{12k} - m_{11k}$, and $p_{\cdot 2k}$ for the Five-college Data

COLLEGE	INTERCEPTS m_{11k}	SLOPES $m_{12k} - m_{11k}$	PROPORTIONS DISAPPROVING OF GETTING DRUNK $p_{\cdot 2k}$
1	0.731	−0.609	0.647
2	0.683	−0.572	0.481
3	0.558	−0.478	0.312
4	0.425	−0.385	0.158
5	0.360	−0.350	0.051

the same model in terms of conditional column proportions as functions of the marginal proportion $p_{\cdot 2k}$. Thus the two separate equations are

$$m_{11k} = b_0 + b_2 p_{\cdot 2k} + u_k \tag{5.20}$$

$$m_{12k} - m_{11k} = b_1 + b_3 p_{\cdot 2k} + v_k \tag{5.21}$$

The observed values of the dependent and independent variables in Eqs. (5.20) and (5.21) for the five-college data contained in Table 5.3 are presented in Table 5.5. Comparing the values of the within-group intercepts and slopes in Table 5.5 suggests why it would be a mistake to assume that the relationship between deviance and tolerance observed at the population level would necessarily hold for individual colleges. Note that the slope $m_{12k} - m_{11k}$ in college 1 is equal to -0.609, while the slope in college 5 is -0.350.

The estimates for the coefficients in Eqs. (5.20) and (5.21) are

$$d_{0k} = m_{11k} = 0.832 - 0.541\bar{x}_k + u_k, \qquad R^2 = 0.97 \tag{5.22}$$

$$d_{1k} = m_{12k} - m_{11k} = -0.677 + 0.383\bar{x}_k + v_k, \qquad R^2 = 0.96 \tag{5.23}$$

In the college example, $b_1 = -0.677$ indicates the individual-level effect of disapproval on getting drunk. The estimate $b_2 = -0.541$ indicates the group effect of disapproval levels in colleges, and the estimate $b_3 = 0.383$ indicates an interaction effect wherein the relationship between individual-level disapproval and getting drunk diminishes as the level of disapproval increases.

Strictly speaking, since the sample sizes and variances of X are not the same in each college, estimates for Eqs. (5.20) and (5.21) ought to be obtained from the weighted regression procedure described in Appendix B. In this instance, the weighted estimates are not significantly different.

With Y and X defined as dummy variables, a second way to estimate the coefficients is through the single multiple regression specified in Chapter 3. With 2×2 table notation, the single equation is obtained by substituting the m's from Eqs. (5.20) and (5.21) into Eq. (5.15). Thus,

$$y_{ik} = b_0 + b_1 x_{ik} + b_2 p_{\cdot 2k} + b_3 x_{ik} p_{\cdot 2k} + (f_{ik} + u_k + v_k x_{ik}) \tag{5.24}$$

Mechanically there are two ways to obtain estimates for the single equation. If the number of cases in each group (table) is relatively small and there are relatively few groups, it may be expedient to actually construct dummy variables for N cases which reflect the distributions of X and Y in each table. For example, from the first college in Table 5.3 there would be 561 cases where $X = 0$ and $Y = 1$, 15 cases where $X = 1$ and $Y = 1$, 206 cases where $X = 0$ and $Y = 0$, and 108 cases where $X = 1$ and $Y = 0$. For $p_{\cdot 2k}$ each individual is assigned the value of the proportion ($p_{\cdot 2k}$) corresponding to his or her college. The variable $x_{ik}p_{\cdot 2k}$ could then be constructed on the computer by multiplying x_{ik} and $p_{\cdot 2k}$. Any standard multiple-regression computer routine would then provide the desired estimates.

This procedure is quite impractical and expensive when the number of cases is large, as in the case of census data. An alternative is to use the means, standard deviations, and correlation coefficients as inputs for the regression computer program. Details of this procedure are contained in Appendix E.

5.1.3. An Alternative Way to Visualize the Contextual Analysis of 2 × 2 Contingency Tables

Another way to visualize contextual analysis for 2 × 2 data is obtained by focusing on Eqs. (5.20) and (5.21) and by solving for m_{12k} in Eq. (5.21). The result is a formal statement of the Davis technique for contextual analysis introduced in Section 1.6.4. Substituting m_{11k} in Eq. (5.20) into Eq. (5.21),

$$m_{12k} - (b_0 + b_2 p_{\cdot 2k} + u_k) = b_1 + b_3 p_{\cdot 2k} + v_k \tag{5.25}$$

Rearranging terms gives

$$m_{12k} = b_0 + b_1 + b_2 p_{\cdot 2k} + b_3 p_{\cdot 2k} + u_k + v_k \tag{5.26}$$

Collecting terms gives

$$m_{12k} = (b_0 + b_1) + (b_2 + b_3) p_{\cdot 2k} + (u_k + v_k) \tag{5.27}$$

It can now be observed in Eq. (5.20) and in Eq. (5.27) that the within-group relationships are expressed in terms of a linear relationship between the conditional proportions (m_{11k} and m_{12k}) and the marginal proportions $p_{\cdot 2k}$, with residuals. For the substantive example, the model specifies that the probabilities of individuals getting drunk change with the proportions of disapproving individuals in the college settings.

In a plot of the column conditional proportions m_{11k} and m_{12k} against the group variable $p_{\cdot 2k}$ the individual, group, and interaction effects show up in a distinctive way. The plot illustrated in Figure 5.1 shows the two lines defined by Eqs. (5.20) and (5.27), leaving out the residuals. In the general case the presence of the interaction effect is indicated when b_3 is different from zero. In that case the two lines have different slopes representing the differences in conditional proportions across the tables as a function of the margin. When no interaction effect is present, the two

Figure 5.1
Relationship between Column Conditional Proportions m_{11} ***and*** m_{12} ***and Group Variable*** $p_{\cdot 2k}$ ***in the Presence of Individual, Group, and Interaction Effects***

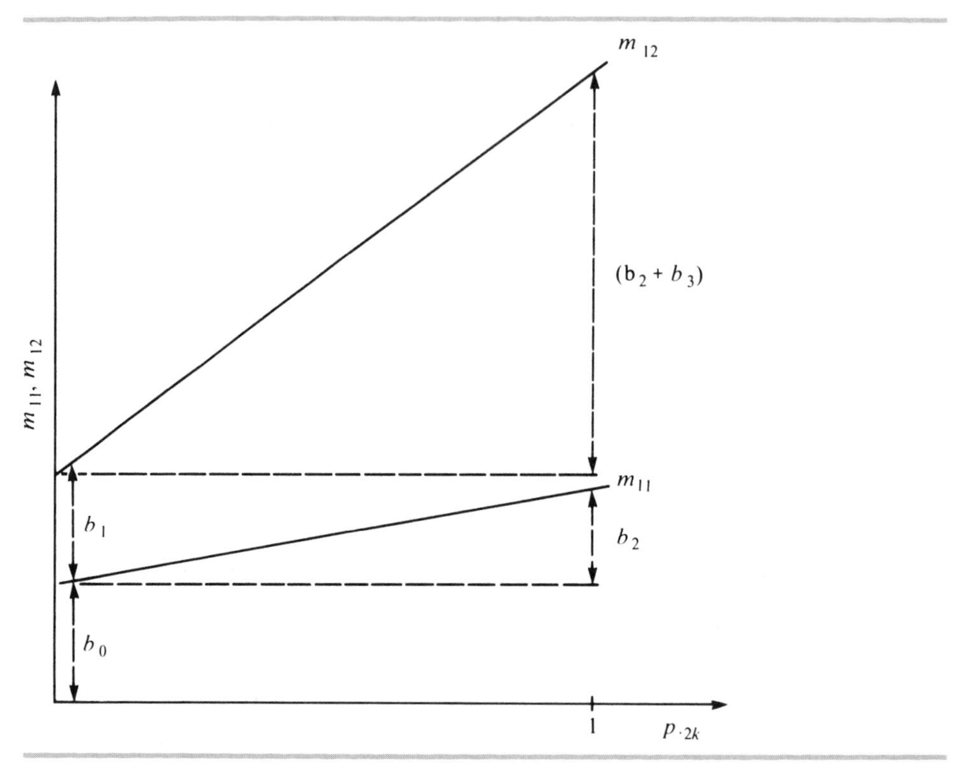

lines will be parallel, and the differences between conditional proportions across the tables will be constant, except for random variation. The group effect is present when the two lines have nonzero slopes, or more specifically when b_2 is different from zero. Finally, the individual-level effect of X is present when b_1 is different from zero, which occurs when the two lines have different intercepts.

For the five-college example the following estimates for Eqs. (5.20) and (5.27) are obtained:

$$m_{11k} = 0.832 - 0.541p_{\cdot 2k} + u_k \quad (5.28)$$

$$m_{12k} = 0.155 - 0.158p_{\cdot 2k} + (u_k + v_k) \quad (5.29)$$

These results are illustrated graphically in Figure 5.2 which shows a difference in the intercepts of the two lines. This indicates the presence of an individual effect on getting drunk associated with personal disapproval. The fact that the line associated with disapproving individuals (m_{11k}) is below the line associated with approving individuals (m_{12k}) indicates that disapproving individuals are less likely than approving individuals to get drunk. The slope of the line associated with persons who do not

Figure 5.2
Conditional Proportions as a Function of the Proportions of Disapproving Individuals by College

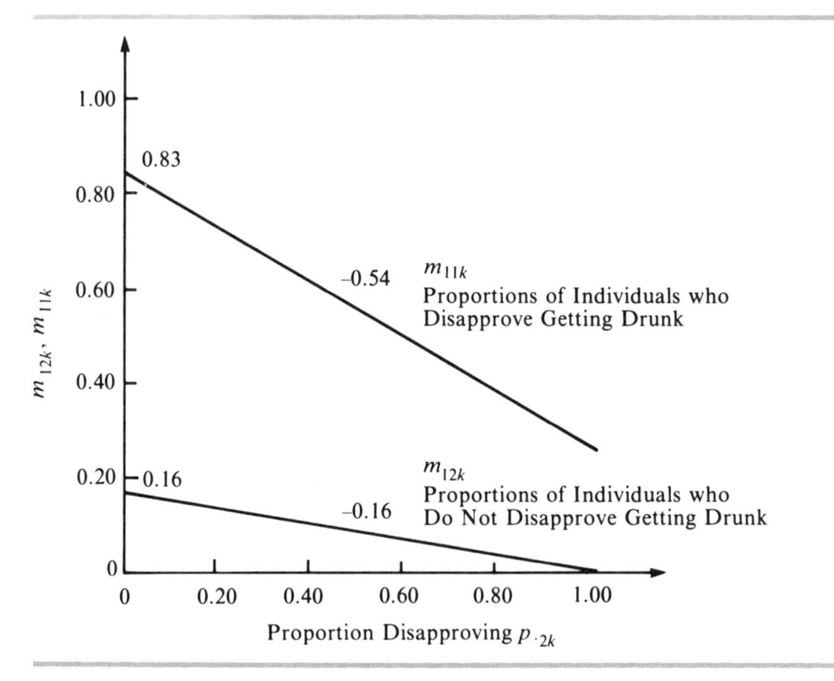

disapprove is different from zero and negative in direction. This indicates the presence of a group effect operating in a negative direction. That is, the odds of approving individuals getting drunk decrease as the college environment gets stricter. The slope of the line associated with disapproving individuals is also negative but less steep than for the approving individuals. This means that persons who disapprove are less affected by the normative climate than are persons who do not disapprove. The difference in these slopes reflects the presence of a cross-level interaction effect.

5.1.4. Measuring and Interpreting Multilevel Effects in 2 × 2 Contingency Tables

Estimates of the coefficients obtained above for the five-college example indicate the presence of individual, group, and cross-level interaction effects. However, because of multicollinearity in the explanatory variables, there remains the problem of interpreting the results in terms of the relative importance of the three variables in explaining the dependent variable. A common practice is to ignore the problem and evaluate explained variation in a stepwise fashion. For example, the individual-level variable may be considered first, in which case it explains 33.6 percent of the variation in Y. If the group variable is included, the model explains 35.7 percent of the variation. Thus the group variable contributes about 2 percent to the explained variation.

When the interaction variable is added, the model explains 36.2 percent of the variation in Y or only 0.5 percent over and above the individual and group variables.

Given high correlations between the explanatory variables, this interpretation must be considered arbitrary in the sense that it depends on the order in which the variables are considered. This procedure almost guarantees that the second variable will look substantially less important than the first, and the third will appear trivial in terms of its contribution. For example, when the group variable is considered first, it accounts for 14.3 percent of the total variation in the dependent variable. The interaction variable adds less than 1 percent in the last position, but it explains 30 percent in the first position.

To deal with this problem, the centering method introduced in Chapter 4 is applied. To get the a coefficients for the centered model with 2 × 2 table data, use

$$d_{0k} = a_0 + a_2(p_{\cdot 2k} - p_{\cdot 2}) + u_k \tag{5.30}$$

$$d_{1k} = a_1 + a_3(p_{\cdot 2k} - p_{\cdot 2}) + v_k \tag{5.31}$$

where $d_{0k} = m_{11k}$ and $d_{1k} = m_{12k} - m_{11k}$. For the college data this gives

$$\begin{aligned} a_0 &= 0.571 && \text{constant} \\ a_1 &= -0.492 && \text{individual effect} \\ a_2 &= -0.541 && \text{group effect} \\ a_3 &= 0.383 && \text{interaction effect} \end{aligned} \tag{5.32}$$

The formulas in Table 4.4 can now be used to compute the sums of squares contributed to the explained variation of the dependent variable by each of the variables. Procedures for obtaining the necessary sums of squares directly from contingency tables are contained in Appendix E. Except for one quantity, all the information needed to compute these sums is readily available in Table 5.3. This quantity is the within-group sums of squares $\sum_i \sum_k (x_{ik} - \bar{x}_k)^2$ which is obtained from Eq. (5.18). For example, this sum for college 1 in Table 5.3 is

$$\begin{aligned} \sum(x_{i1} - \bar{x}_1)^2 &= s_1^2 n_1 = p_{\cdot 21} p_{\cdot 11} n_1 \\ &= (0.138)(0.862)(890) = 105.871 \end{aligned} \tag{5.33}$$

The results of the centering procedure are given in Table 5.6. Observe that the three variables together explain 30.3 percent of the variation in the dependent variable. The individual level accounts for 77.8 percent of the explained variation. The group level accounts for 20.7 percent, and the interaction accounts for 1.5 percent.

Results of the centering procedure for this example indicate that the group variable regarding norms about getting drunk is considerably more important to the explanation of getting drunk than was indicated by the previous stepwise procedure which arbitrarily considered the individual variable first. On the other hand, the results indicate that the group variable is considerably less important than the individual variable in a stepwise analysis when it is considered first.

Table 5.6
Sum-of-squares Table for the Five-college Contextual Analysis after Centering

	Explained variation	Unexplained variation	Total explained and unexplained
Individual level	262.73 77.8* 25.4 23.6	771.07 99.3 74.6 69.2	1033.8 92.8
Group level	70.06 20.7 95.3 6.3	3.45 0.4 4.7 0.3	73.51 6.6
Interaction level	4.91 1.5 72.1 0.4	1.90 0.3 27.9 0.2	6.81 0.6
Total	337.70 30.3	776.42 69.7	1114.12 100.0

* Except for the marginal totals, percents are given first by column, then by row, and then corner.

5.2
Contextual Analysis with Multiple-category Contingency Tables

Contingency tables often have more than two rows and columns. Suppose we were analyzing the impact of political party identification on individual voting behavior. In general it could be predicted that individuals would vote for the candidate closest to their party affiliation. The contextual hypothesis might be that, within districts where Democrats were in the majority, individuals would be more likely to vote for a Democratic candidate than in districts where the Republicans were in the majority, that is, regardless of personal party affiliation. However, the partisan identification "independent" and the relative numbers of independents in a voting district might be important in choosing candidates.

There also might be more than two categories of the dependent variable. We might be trying to explain an individual-level choice between two options only to find, as is often the case, that there is really a meaningful third choice. In this instance it might be the alternative to not vote at all.

When this occurs, a variable cannot be represented by a single dummy variable, and the relationship between the two individual-level variables cannot be studied using a simple-regression analysis. We do not have an intercept and a slope to define the three types of effects. But it is possible to generalize the treatment of contingency tables by introducing several dummy variables and making use of several multiple-regression analyses.

5.2.1. Notation for the Multiple-category Case

In the general case we have K contingency tables, each with r rows and c columns. The various equations get cumbersome to write, and without much loss of generality the presentation is restricted to tables with three rows and three columns. The extension to larger tables is straightforward.

Table 5.7 shows a 3×3 contingency table for the kth group. The table also shows how the variable Y is represented by two dummy variables Y_1 and Y_2. Each of the $n_{1 \cdot k}$ individuals belonging to the first category of Y are assigned the value of 1 on the variable Y_1 and 0 on the variable Y_2. Similarly, each of the $n_{2 \cdot k}$ individuals in category 2 are assigned the value of 0 on Y_1 and 1 on Y_2. Finally, the $n_{3 \cdot k}$ individuals in category 3 are assigned the value of 0 on both Y_1 and Y_2. Similarly, the variable X is represented by the two dummy variables X_1 and X_2. All individuals in category 1 are assigned the value of 1 on X_1 and 0 on X_2, all individuals in category 2 are assigned the value of 0 on X_1 and 1 on X_2, and all individuals in the third category are assigned the value of 0 on both X_1 and X_2.

Table 5.7
Observed Frequencies and Allocation of Values to Dummy Variables for a 3 × 3 Contingency Table

			X			Dummy variables	
		1	2	3	Total	Y_1	Y_2
Y	1	n_{11k}	n_{12k}	n_{13k}	$n_{1 \cdot k}$	1	0
	2	n_{21k}	n_{22k}	n_{23k}	$n_{2 \cdot k}$	0	1
	3	n_{31k}	n_{32k}	n_{33k}	$n_{3 \cdot k}$	0	0
Total		$n_{\cdot 1k}$	$n_{\cdot 2k}$	$n_{\cdot 3k}$	n_k		
Dummy variables	X_1	1	0	0			
	X_2	0	1	0			

In the kth group the relationship between each of the two dummy variables representing Y and the dummy variables representing X can be expressed in the equations

$$\begin{aligned} y_{1ik} &= d_{10k} + d_{11k} x_{1ik} + d_{12k} x_{2ik} + f_{1ik} \\ y_{2ik} &= d_{20k} + d_{21k} x_{1ik} + d_{22k} x_{2ik} + f_{2ik} \end{aligned} \tag{5.34}$$

The d's that minimize the two residual sums of squares are functions of the conditional-column proportions and can be written

$$\begin{aligned} d_{10k} &= m_{13k} \\ d_{11k} &= m_{11k} - m_{13k} \\ d_{12k} &= m_{12k} - m_{13k} \\ d_{20k} &= m_{23k} \\ d_{21k} &= m_{21k} - m_{23k} \\ d_{22k} &= m_{22k} - m_{23k} \end{aligned} \tag{5.35}$$

Each proportion m is obtained by dividing the corresponding cell frequency by the total for the column in which the cell is located, that is,

$$m_{ijk} = \frac{n_{ijk}}{n_{\cdot jk}} \tag{5.36}$$

The separate model equations specify that the d's are functions of the means of the two variables X_1 and X_2. From the way these dummy variables are defined, it follows that the mean of X_1 equals the proportions $p_{\cdot 1k}$ and the mean of X_2 equals the proportion $p_{\cdot 2k}$. With that the separate model equations are written:

$$\begin{aligned} d_{10k} &= b_{100} + b_{101}p_{\cdot 1k} + b_{102}p_{\cdot 2k} + u_{10k} \\ d_{11k} &= b_{110} + b_{111}p_{\cdot 1k} + b_{112}p_{\cdot 2k} + u_{11k} \\ d_{12k} &= b_{120} + b_{121}p_{\cdot 1k} + b_{122}p_{\cdot 2k} + u_{12k} \\ d_{20k} &= b_{200} + b_{201}p_{\cdot 1k} + b_{202}p_{\cdot 2k} + u_{20k} \\ d_{21k} &= b_{210} + b_{211}p_{\cdot 1k} + b_{212}p_{\cdot 2k} + u_{21k} \\ d_{22k} &= b_{220} + b_{221}p_{\cdot 1k} + b_{222}p_{\cdot 2k} + u_{22k} \end{aligned} \tag{5.37}$$

The notation is now more complex, but there is a definite pattern to it. For the d's the first subscript refers to the Y variable, the second refers to the X variable, and the third refers to the group. The b's also have three subscripts, where the first refers to the Y variable, the second refers to the X variable, and the third refers to the mean of the X variable. The b values in each particular equation minimize the residual sum of squares for that equation across all the groups.

To identify the various effects, the d's from Eq. (5.37) are substituted into Eq. (5.34). The first equation becomes

$$\begin{aligned} y_{1ik} = \; & b_{100} && \text{constant} \\ & + b_{110}x_{1ik} + b_{120}x_{2ik} && \text{individual} \\ & + b_{101}p_{\cdot 1k} + b_{102}p_{\cdot 2k} && \text{group} \\ & + b_{111}x_{1ik}p_{\cdot 1k} + b_{112}x_{1ik}p_{\cdot 2k} && \text{interaction} \\ & + b_{121}x_{2ik}p_{\cdot 1k} + b_{122}x_{2ik}p_{\cdot 2k} && \\ & + f_{ik} + u_{10k} + u_{11k}x_{1ik} + u_{12k}x_{2ik} && \text{residuals} \end{aligned} \tag{5.38}$$

Eq. (5.38) is grouped into terms that measure the individual, group, and interaction effects. A similar equation is obtained for Y_2, and it becomes

$$
\begin{aligned}
y_{2ik} = {} & b_{200} && \text{constant} \\
& + b_{210}x_{1ik} + b_{220}x_{2ik} && \text{individual} \\
& + b_{201}p_{\cdot 1k} + b_{202}p_{\cdot 2k} && \text{group} \\
& + b_{211}x_{1ik}p_{\cdot 1k} + b_{212}x_{1ik}p_{\cdot 2k} && \text{interaction} \\
& + b_{221}x_{2ik}p_{\cdot 1k} + b_{222}x_{2ik}p_{\cdot 2k} && \\
& + f_{2ik} + u_{20k} + u_{21k}x_{1ik} + u_{22k}x_{2ik} && \text{residuals}
\end{aligned}
\tag{5.39}
$$

These last two equations help us interpret the various regression coefficients obtained from Eq. (5.37). Because they are the coefficients for the individual and group variables, respectively, the four coefficients b_{110}, b_{120}, b_{210}, and b_{220} measure the individual-level effect of X, and the four coefficients b_{101}, b_{102}, b_{201}, and b_{202} measure the group effect. The remaining coefficients are for the eight product variables, and they measure the interaction effect.

5.2.2. The Alternative Way to Visualize Effects in Multiple-category Tables

The discussion of the 2×2 case shows how the presence and absence of effects could be visually observed in a plot. Under special conditions this method can also be used with larger contingency tables. These conditions exist when we are able to decide which marginal proportion is of most substantive interest and most likely to be the major source of compositional effects. If there are more than two categories of the dependent variable, we would also have to decide which of these categories is of the greatest interest. This would amount to the assumption that a number of the parameters in the set of equations under Eq. (5.37) are equal to zero. This assumption could, of course, be supported by finding that the corresponding estimates did turn out to be zero or negligible.

For illustration, suppose that the analysis objective is to explain the outcome of an initiative concerning aid to private schools. The possible outcomes are a vote for, a vote against, and a no show. The individual-level variable of interest is the religious affiliation of the respondent: Catholic, Protestant, and other. The group variable of interest is the religious composition of the respondent's voting district. The data are set up as in Table 5.8.

We decide that the no show group is no different in relevant respects than the general population of voters, and therefore leave it out of consideration. We also decide that the relative dominance of the Catholic church in the voter's district is the major source of context effects. This narrows the model of Eq. (5.37) to the following:

$$
\begin{aligned}
d_{10k} &= b_{100} + b_{101}p_{\cdot 1k} + u_{10k} \\
d_{11k} &= b_{110} + b_{111}p_{\cdot 1k} + u_{11k} \\
d_{12k} &= b_{120} + b_{121}p_{\cdot 1k} + u_{12k}
\end{aligned}
\tag{5.40}
$$

Table 5.8
Hypothetical Example for Contextual Analysis with Larger Tables (with Column Conditional Proportions and Marginal Proportions)

	Catholic	Protestant	Other	
Voted for	m_{11k}	m_{12k}	m_{13k}	$p_{1\cdot k}$
Voted against	m_{21k}	m_{22k}	m_{23k}	$p_{2\cdot k}$
Did not vote	m_{31k}	m_{32k}	m_{33k}	$p_{3\cdot k}$
	$p_{\cdot 1k}$	$p_{\cdot 2k}$	$p_{\cdot 3k}$	1.00

Now the separate equations can be rewritten in terms of the column conditional proportions m_{11k}, m_{12k}, and m_{13k}. This gives

$$\begin{aligned} m_{13k} &= b_{100} + b_{101}p_{\cdot 1k} + u_{13k} \\ m_{12k} &= (b_{120} + b_{100}) + (b_{121} + b_{101})p_{\cdot 1k} + (u_{12k} + u_{10k}) \\ m_{11k} &= (b_{110} + b_{100}) + (b_{111} + b_{101})p_{\cdot 1k} + (u_{11k} + u_{10k}) \end{aligned} \tag{5.41}$$

These equations can be plotted as before in the case of the 2×2 data, except that we can now observe the effects of the context variable on three groups. Just to illustrate, we might find the (hypothetical) situation pictured in Figure 5.3.

Figure 5.3
Relationships between Three Conditional Proportions and One Marginal Proportion (for hypothetical example)

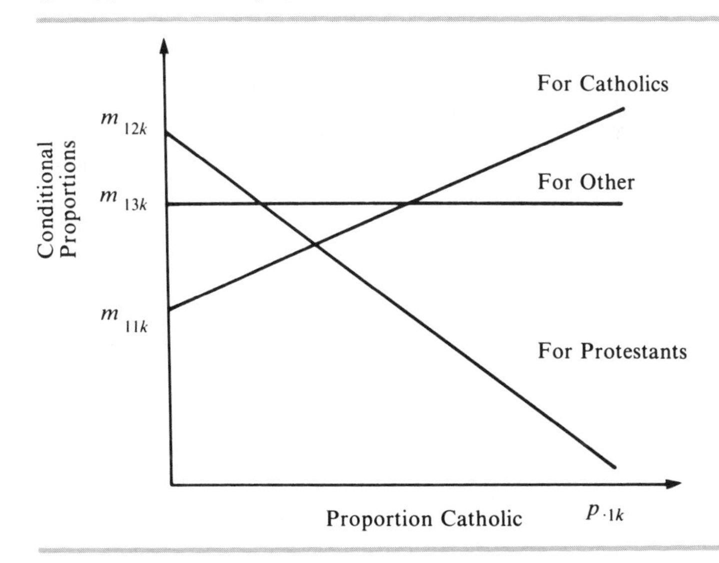

The utility of a graph such as Figure 5.3 is that we can easily visualize the relationships between groups at various levels of the context variable. For example, when the religious composition of the district is predominantly non-Catholic, support for aid is highest among Protestants and lowest among Catholics. When the composition of the district is predominantly Catholic, the reverse is true. This might lead to speculation that the Protestants in these districts are reacting to the greater salience of the parochial school issue. It could also be observed that the population that is neither Protestant nor Catholic appears to be stable in its support for aid and unaffected by the relative strength of the Catholic church. It might be surmised that this group supports or opposes aid on grounds other than the parochial school issue. This puts them less in support of aid than the Protestants and more in support than the Catholics when the district is non-Catholic and the reverse when the district is Catholic.

5.3
Summary

This chapter shows that the basic model of contextual analysis for continuous variables examined in Chapters 3 and 4 can be applied to nominal variables through the use of dummy variables. Considerable attention is directed to the relationships between the terms and concepts of contingency tables and the terms and operations of regression analysis. Out of this comes the formalization of an older but complementary way to visualize individual, group, and interaction effects with categorical variables. Rather than viewing within-group slopes and intercepts as functions of group means, this approach views the conditional proportions for each category of the individual-level variable as functions of marginal proportions.

By identifying the relationships between regression operations and data already summarized in table form, a complete contextual analysis could be carried out directly from tables, including the estimation and evaluation of effects. The implication is that it is not necessary to prepare a data set consisting of observations for every individual in the study. This opens the way to analyze data, such as the U.S. census, which cover large populations in tabular form.

Contextual Analysis for the Treatment of Nonlinear Context Effects and More Than One Set of Explanatory Variables

Early contextual analyses were properly criticized on the grounds that they oversimplified reality by considering only one individual-level variable.[1] They were also criticized for considering only linear relationships between variables. This is not a valid criticism of contextual analysis *per se*. Nothing prevents extending the basic model to include additional variables and the use of nonlinear functions any more than in conventional multivariate analysis. The purpose of this chapter is to examine these extensions.

6.1 *Nonlinear Context Effects in Contextual Analysis*

6.1.1. The Case for Continuous Variables

This important extension of the basic model of contextual analysis involves the presence of group or interaction effects which increase or decrease nonlinearly as a function of the group-mean variable $\bar{X}$. For example, in the hypothetical four-city study of deviance and social norms (Table 3.1) there might have been reason to expect deviance to increase rapidly with tolerance levels in the lower ranges of that context variable but not in the higher. Such a phenomenon is generally visible in a plot of the observed values of the intercepts d_{0k} and the city means $\bar{x}_k$. This theory could be explored by dividing the groups at the overall mean of X and performing the

[1] These criticisms of early contextual analyses are discussed along with other issues in Chapter 13.

Figure 6.1
Polynomial Curve Relating Intercepts and Group Means in the Four-city Data (without Residuals)

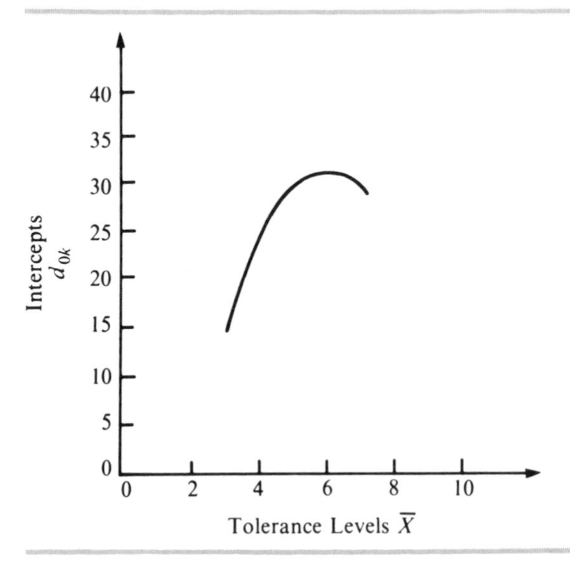

regression in Eq. (3.6) for each group. Although the four-city example is somewhat unrealistic because of the small number of groups (cities), this procedure results in a slope of 6.0 for the cities with lower levels of tolerance and a slope of -1.25 at the higher levels.[2]

A better way to study this phenomenon is to fit a curve rather than a line to the observations relating the intercepts and the means. This curve can take many forms, but a common form is expressed in the polynomial containing a squared term as in

$$d_{0k} = b_0 + b_2\bar{x}_k + b_4\bar{x}_k^2 + u_k \tag{6.1}$$

The coefficient b_4 associated with the squared term $\bar{x}_k^2$ provides a measure of second-degree group effects. The coefficients for the four-city data contained in Table 3.1 are

$$d_{0k} = -34.56 + 21.62\bar{x}_k - 1.81\bar{x}_k^2 + u_k, \qquad R^2 = 0.937 \tag{6.2}$$

The curve reflected in Eq. (6.2) is shown in Figure 6.1. The magnitude of the nonlinear effects can also be inferred from the change in R^2 after introducing this variable. In this case the value of R^2 without the squared variable is 0.613 (see Table 3.1). With the squared variable included, R^2 equals 0.937. The substantial difference between the two estimates confirms the existence of a second-degree group effect.

Nonlinear relationships may also emerge in contextual analysis in the relationship connecting the within-group slopes and means. In the four-city data the within-group relationships between tolerance and deviance could be weak in the lower

[2] For the sake of discussion, the four cities were divided into two groups of two cities each, even though two of the cities had the same value on levels of tolerance $\bar{X}$.

ranges of the context variable but strong in the higher levels. This theory suggests that where the climate is generally unaccepting of deviant behavior, there is not much difference between tolerant and unaccepting individuals with respect to deviant behavior. But where the norms are generally accepting, there are substantial differences between accepting and unaccepting individuals.

This phenomenon can be studied by introducing a squared term to Eq. (3.7) to detect second-degree interaction effects. Thus,

$$d_{1k} = b_1 + b_3\bar{x}_k + b_5\bar{x}_k^2 + v_k \tag{6.3}$$

The estimate of b_5 in the four-city study equals zero, indicating the absence of this kind of effect.

The single-equation alternative for estimating effects can be obtained as above by substituting for d_{0k} and d_{1k} from the separate equations [Eqs. (6.1) and (6.3)] into Eq. (3.3). This gives

$$y_{ik} = b_0 + b_1 x_{ik} + b_2\bar{x}_k + b_3 x_{ik}\bar{x}_k + b_4\bar{x}_k^2 + b_5 x_{ik}\bar{x}_k^2 + (u_k + v_k x_{ik} + f_{ik}) \tag{6.4}$$

where $u_k + v_k x_{ik} + f_{ik} = e_{ik}$. Measures of the various effects including second-degree group and interaction effects are provided by the corresponding regression coefficients.

6.1.2. The Impact of Nonlinear Context Effects on Separate- and Single-equation Estimates

The example demonstrates how extending the basic model of contextual analysis to detect nonlinear context effects may be necessary for substantive reasons. There is, however, another important reason for being concerned about these effects. When nonlinear effects are present but ignored in the model specification, the separate- and single-equation procedures will produce *different* estimates. The coefficients from the separate- and single-equation procedures, without and with the squared terms, are summarized in Table 6.1. In the case of the individual effect, the coefficients are of opposite signs. When nonlinear effects are explicitly specified in the model, the estimates are identical. The reasons for this are discussed in Appendix F. A discrepancy between the separate- and single-equation results for the basic model should be a signal that the model is not correctly specified. Such a discrepancy should lead to further substantive insights about the operation of contextual factors.

6.1.3. Evaluating Effects in Models Containing Nonlinear Context Effects

Evaluating the contributions of the variables in the analysis is complicated by the presence of multicollinearity. Since the variable $\bar{X}^2$ is a direct transformation of the other variables, it will be correlated with them. In the example the correlation between the means and the squared means is 0.99. Table 6.2 presents the full sum-of-squares table for the four-city data where the individual-level variable is considered first.

Table 6.1
Regression Coefficients from Separate- and Single-equation Procedures without and with Polynomial Variables

	CONSTANT b_0	INDIVIDUAL EFFECT b_1	GROUP EFFECT b_2	INTERACTION EFFECT b_3	POLYNOMIAL GROUP EFFECT b_4	POLYNOMIAL INTERACTION EFFECT b_5
Without polynomial variables						
Separate-equation procedure	7.13	−1.25	3.50	1.25	—	—
Single-equation procedure	−13.72	3.28	8.03	0.34	—	—
With polynomial variables						
Separate-equation procedure	−34.56	−1.25	21.62	1.25	−1.81	0.00
Single-equation procedure	−34.56	−1.25	21.62	1.25	−1.81	0.00

Table 6.2
Sum-of-squares Table with Nonlinear Model for the Four-city Data

	Explained by specified variables	Unexplained by specified variables	Total explained and unexplained variation
Individual-level variable considered first	6241.0	2225.0	8466.0
Group, polynomial group, and interaction variables together	1641.3	76.5	1717.8
Total explained variation	7882.3	2301.5	10183.8

In this position it accounts for 6241.0/10,183.8 = 61.3 percent of the total variation in Y. The other variables together, including the squared-means variable, explain 16.1 percent when they are introduced after the individual-level variable. Because the explanatory variables are correlated, very different results are obtained when the other variables (together) are considered first. Although not shown here, the group, second-degree group, and interaction variables together explain virtually all 77.4 percent of the total variation explained by the model containing all the variables.

As explained in Chapter 4, a complete interpretation of the full sum-of-squares table is carried out by computing percentages to the columns and rows as well as to the corner. For example, the model containing all the specified variables (including the squared variable) explains 7882.3/10,183.8 = 77.4 percent of the total variation. This compares to 76.4 percent for the model excluding the squared variable. Similarly, the three context variables together account for 20.8 percent of the explained variation. Compare this to 19.8 percent in the model which did not include the second-degree group variable. Including the squared variable reduces the amount of unexplained variation attributed to the three context-related variables to 76.5/2301.5 = 3.3 percent. This leaves 96.7 percent of the total unexplained variation attributed to unspecified individual-level variables.

In the discussion of the centering procedure (Chapter 4), it is observed that the stepwise procedure may exaggerate the importance of the specified variables, depending on which variable is considered first. To help evaluate the unique contributions of the variables, the centering operations specified for the basic model are extended to include nonlinear context effects. For the separate equations this involves measuring the variables in Eqs. (6.1) and (6.3) around their respective overall means as in

$$d_{0k} = a_0 + a_2(\bar{x}_k - \bar{x}) + a_4(\bar{x}_k^2 - \overline{\bar{x}^2}) + u_k \tag{6.5}$$

$$d_{1k} = a_1 + a_3(\bar{x}_k - \bar{x}) + a_5(\bar{x}_k^2 - \overline{\bar{x}^2}) + v_k \tag{6.6}$$

The coefficients have been changed to *a*'s, indicating that they refer to the centered model.

The concern for nonlinear effects leads only to a change in the equations involving the intercepts and slopes and not the individual-level equation [Eq. (3.3)]. The procedures for adjusting Y therefore remain the same, as described in Section 4.3. For the single-equation procedure,

$$y'_{ik} = d_{0k} + d_{1k}(x_{ik} - \bar{x}_k) + f_{ik} \tag{6.7}$$

With one exception the centering procedure removes the correlations between the explanatory variables. The exception is that the correlation between $\bar{X}$ and $\overline{X^2}$ is not removed. The variation between these two variables therefore cannot be broken into two unique parts. This is not a serious problem because both variables refer to group effects. They can be treated together in the sum-of-squares table. The explained variation can be divided into three unique parts corresponding to the individual, group, and interaction variables. The results of this analysis are summarized in Table 6.3.

As in Chapter 4, the results of the analysis can be summarized by computing proportions by row, column, and corner. The individual-level variable accounts for 800.0/1502.3 = 53.3 percent of the total explained variation. The group variable and the squared-group variable together account for 40.1 percent, and the interaction variable accounts for 6.7 percent. In contrast to the uncentered model, the centered model traces much more of the explanation of the dependent variable (almost half) to context-related variables. On the other hand, very little of the unexplained variation is accounted for by the context variables. This suggests again

Table 6.3
Sum-of-squares Table with Centered Nonlinear Model for the Four-city Data

	Explained by specified variables	Unexplained by specified variables	Total explained and unexplained variation
Individual-level variable	800.0	2225.0	3025.0
First and second order group variables	602.3	40.5	642.8
Interaction variable	100.0	16.0	116.0
Total explained variation	1502.3	2281.5	3783.8

that the search for additional variables ought to concentrate on other unspecified individual-level variables.

6.1.4. An Illustration of Nonlinear Context Effects with Dichotomous Individual-level Variables

Although individual-level variables may be dichotomous, the means ($\bar{x}_k = p._{2k}$), the intercepts ($d_{0k} = m_{11k}$), and the slopes ($d_{1k} = m_{12k} - m_{11k}$) are interval variables and can therefore be related in nonlinear ways. With appropriate dummy-variable transformations of the individual-level variables, the principles and procedures involving nonlinear context effects described above generalize completely. When applied to the previous example based on Bower's study of college norms and getting drunk, only linear effects are detected.

However, the effects are striking when the procedures are applied to the data which Robinson (1950) used in his famous discussion of the ecological fallacy. These data consisted of illiteracy rates and proportions of foreign-born, by state, as reported in the 1930 census of the population. These variables are recorded in the study as dichotomous categories. They are treated here as dummy variables, with the category foreign-born and the category illiterate assigned 1's and the other categories 0's. Accordingly, $\bar{X}$ represents the state proportions of foreign-born.

When the possibility of nonlinear context effects are ignored in a contextual analysis of these data, the following results are obtained from the separate equations:

$$d_{0k} = 0.08 - 0.33\bar{x}_k + u_k, \qquad R^2 = 0.467 \tag{6.8}$$

$$d_{1k} = -0.01 + 0.42\bar{x}_k + v_k, \qquad R^2 = 0.418 \tag{6.9}$$

Because the coefficient for the individual effect is close to zero, there is essentially no (pure) individual effect of being foreign-born on being illiterate. As the composition of state populations indicates proportionately greater numbers of foreign-born, the odds of being illiterate decrease for native born. This is a group effect.

Substantively speaking, at the time of the 1930 census of the population, states with larger proportions of foreign-born also tended to be states with better quality or more available public education. It is likely that the group effect is attributable to this phenomenon. There is little difference between the two populations when the proportion of foreign-born is small, but there are increasingly larger differences as the proportion of foreign-born increases. This cross-level interaction effect probably reflects inequalities in access or availability of public education that transcended the overall quality of education within states.

An explicit consideration of nonlinear context effects in the model specification produces these results:

$$d_{0k} = 0.10 - 0.98\bar{x}_k + 2.37\bar{x}_k^2 + u_k, \qquad R^2 = 0.651 \tag{6.10}$$

$$d_{1k} = 0.03 + 0.39\bar{x}_k - 0.45\bar{x}_k^2 + v_k, \qquad R^2 = 0.431 \tag{6.11}$$

Compare the R^2 values in Eqs. (6.10) and (6.11) with the corresponding R^2 values in Eqs. (6.8) and (6.9). The squared variable in Eq. (6.10) significantly increased the proportion of variance explained by that model, but the squared variable in Eq. (6.11) did not add anything significant. The presence of a second-degree group effect is inferred, but not the presence of a second-degree interaction effect. A comparison of Eqs. (6.8) and (6.10) shows that the incorrectly specified model appreciably underestimates the magnitude of group effects. The group effect and the second-degree group effect are of opposite signs. There is a group effect at work in this example, but it decidedly fades as the proportion of foreign-born increases. The phenomenon is observed graphically in Figure 6.2.

Figure 6.2
Scatter Plot of the Relationship between Group Means and Group Intercepts

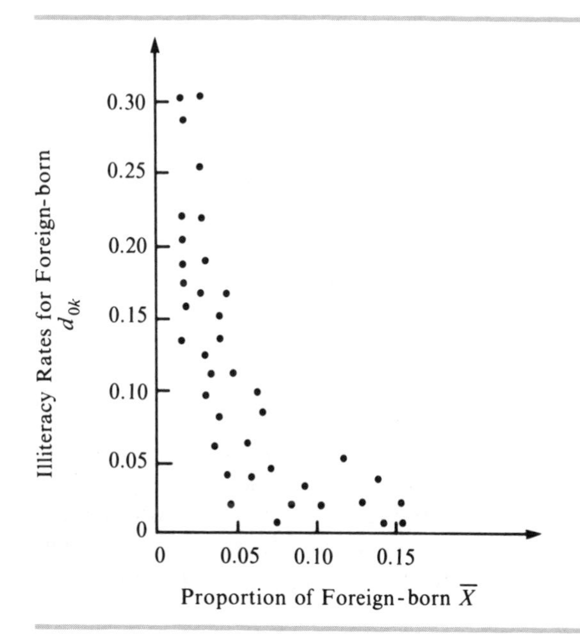

Table 6.4
Regression Coefficients from Separate- and Single-equation Procedures without and with the Polynomial Group Variable for Illiteracy Data

		CONSTANT b_0	INDIVIDUAL EFFECT b_1	GROUP EFFECT b_2	INTERACTION EFFECT b_3	POLYNOMIAL GROUP EFFECT b_4
Without polynomial group variable	Separate-equation procedure	0.08	−0.01	−0.33	0.42	—
	Single-equation procedure	0.10	−0.47	−0.52	2.79	—
With polynomial group variable	Separate-equation procedure	0.10	−0.01	−0.98	0.42	2.37
	Single-equation procedure	0.10	−0.01	−1.00	0.42	2.37

As in the previous example, this case illustrates the importance of nonlinear context effects on the separate and single-equation estimates of effects. Results for the two methods, with and without the squared terms, are summarized in Table 6.4. These results reveal a sharp discrepancy between the two sets of coefficients when the necessary squared term is left out. They also show how this discrepancy virtually disappears when the model includes a squared term.

6.2 *Contextual Analysis with More Than One Set of Explanatory Variables*

6.2.1. Expanding the Basic Model

Theory construction in the social sciences frequently involves more than one individual-level explanatory variable. The basic model of contextual analysis which specified only one individual variable must therefore be expanded to specify other individual variables. The hypothetical four-city study theorized that personal acts of deviance are affected by personal attitudes toward deviance. No one would expect that personal tolerance is the *only* individual characteristic affecting deviant behavior. A more complex theory might involve the role of socioeconomic status (SES).

Since SES is an individual-level variable, it is introduced to Eq. (3.3) as part of the effort to explain Y within groups. This produces the multiple-regression models contained in Eq. (6.12):

$$y_{ik} = d_{0k} + d_{1k}x_{1ik} + d_{2k}x_{2ik} + f_{ik} \tag{6.12}$$

In this illustration X_1 and X_2 represent the variables tolerance and SES.

With the addition of the second individual-level variable, the same concern should arise for context effects associated with its group means as for the first variable. That is, we should be concerned about the possibility that the SES levels in the individual's group (city) also influence deviant behavior, independent of personal SES. Directly extending the logic of the basic model, Eq. (3.6) becomes

$$d_{0k} = b_{00} + b_{01}\bar{x}_{1k} + b_{02}\bar{x}_{2k} + u_{0k} \tag{6.13}$$

The notation is altered to identify the group variables, the measures of effects associated with them, and the residuals. In terms of interpretation, Eq. (6.13) says that we expect group effects, as reflected in different within-group intercepts, to be a function of two group-level variables, tolerance levels and SES levels, plus the residual u_{0k}.

The new source of cross-level interaction effects accompanying the second individual-level variable is left to be specified. Following directly from the basic model, there is a separate equation for each of the within-group partial regression coefficients d_{1k} and d_{2k}. Thus,

$$d_{1k} = b_{10} + b_{11}\bar{x}_{1k} + b_{12}\bar{x}_{2k} + u_{1k} \tag{6.14}$$

$$d_{2k} = b_{20} + b_{21}\bar{x}_{1k} + b_{22}\bar{x}_{2k} + u_{2k} \tag{6.15}$$

Eqs. (6.14) and (6.15) are consistent with the model that cross-level interaction effects, as reflected in differences in the within-group partial slopes, are functions of the two group variables and associated unexplained effects.

The single equation for this extended model is obtained as before. Substituting the d's from Eqs. (6.13), (6.14), and (6.15) into Eq. (6.12), we get

$$\begin{aligned} y_{ik} = {} & b_{00} + b_{10}x_{1ik} + b_{20}x_{2ik} + b_{01}\bar{x}_{1k} + b_{02}\bar{x}_{2k} \\ & + b_{11}x_{1ik}\bar{x}_{1k} + b_{12}x_{1ik}\bar{x}_{2k} + b_{21}x_{2ik}\bar{x}_{1k} + b_{22}x_{2ik}\bar{x}_{2k} \\ & + f_{ik} + u_{0k} + u_{1k}x_{1ik} + u_{2k}x_{2ik} \end{aligned} \tag{6.16}$$

The top line of Eq. (6.16) contains terms for the individual and group effects. The second line contains terms for each possible interaction. The bottom line contains residual terms associated with the individual level, the group level, and the interaction level.

While this model extension is described using interval-level variables, categorical variables can also be handled with the appropriate dummy-variable transformations. These additional variables might include employment status, race, sex, occupation, marital status, and so forth. Group-level variables other than those based on means or proportions also might be necessary to explain the dependent variable. In the four-city study, variables reflecting formal structures such as type of government or constitutional arrangements for the control of deviance might be important.

6.2.2. The Possibility of Interactions on the Same Level

Additional considerations emerge with analyses involving more than a single set of explanatory variables. For one, treating more than one variable at each level raises the possibility of statistical interactions *within* levels. Tolerant or intolerant attitudes, for example, may go with SES to produce deviant behavior. If this interaction is assumed to be multiplicative in form, it could be introduced to Eq. (6.12) in the product term $(x_1 x_2)_{ik}$ with a corresponding coefficient d_{3k} for measuring its effects. We should then consider the possibility that this coefficient, like the others in Eqs. (6.13), (6.14), and (6.15), varies systematically with the two group variables $\bar{X}_1$ and $\bar{X}_2$. With the existence of two group variables in Eqs. (6.13), (6.14), and (6.15) instead of one as in the basic model, another possibility is the more complex contextual environment, with an interaction between levels of tolerance and average SES. This can be treated in the analysis by including the product term $(\bar{x}_1 \bar{x}_2)_k$ in Eqs. (6.13), (6.14), and (6.15).

6.2.3. Evaluating Effects with More Than One Set of Explanatory Variables

Introducing additional variables intensifies the problem of correlated explanatory variables and, thereby, the problem of interpreting the analysis results in terms of the relative importance of the explanatory variables. The basic model establishes that we could expect the individual, group, and interaction variables to be intercorrelated. With two or more predictor variables in each separate equation, we again could anticipate multicollinearity. In the four-city case we could expect the individual-level variables tolerance and SES in Eq. (6.12) to be correlated. We could also expect the group variables tolerance level and average SES in Eqs. (6.13), (6.14), and (6.15) to be correlated. This is the same situation encountered in the basic model, complicated here by more places for multicollinearity to occur. With multicollinearity on the same level of observation, the contribution of the variables to the explanation of variance cannot be partitioned into unique parts associated with each variable. With multicollinearity in the explanatory variables at different levels of observation, we also cannot hope to break up the total explained variance into parts uniquely associated with the individual-level variables considered together, the group variables considered together, and the interaction variables considered together.

Nevertheless, the model involving more than one set of explanatory variables can be summarized by applying a stepwise procedure. One possibility is to summarize the stepwise contributions of *blocks* of variables, rather than of one variable at a time. Here there is a choice. One way is to put all the individual-level variables together in one block, all of the group variables in another, and all the interaction variables in still another block. Alternatively, the explanatory variables might be organized into groups consisting of the natural sets. With two individual-level variables there would be two blocks consisting of an individual-level variable and its group, interaction, and higher degree transformations. After specifying the model, one would again have to decide the order in which to introduce the groups of variables into the analysis.

The variation left unexplained by the explanatory variables can also be divided into meaningful parts. One part will consist of the sums of squares left unexplained by the individual-level variables and the other of the sums left unexplained by the context variables (group and interaction). For a detailed discussion of this, see Appendix G.

6.2.4. The Centering Procedure for More Than One Set of Explanatory Variables

The centering procedure can be applied to evaluate the effects in the case involving more than one set of explanatory variables. Centering removes, at least at crucial points, correlations between explanatory variables. The analyst can then trace influences on a dependent variable to different levels of observation. The essential logic of this procedure is described for the basic model. Dependent and explanatory variables are transformed to shift the within-group regression lines over a single point while preserving the values of the within-group slopes and intercepts. The difference with more than one set of explanatory variables is that we shift planes and hyperplanes, instead of simple lines. For this reason it becomes increasingly difficult to meet all the objectives of centering. Generally speaking, it will not be possible to completely remove the correlations between variables located at the same level of observation. It is possible, however, to remove or significantly reduce correlations between blocks of variables that correspond to levels of observation. The analyst can then make decisive statements about the relative importance of different levels of influence on individual-level behavior.

Generalizing from the centering procedure prescribed for the basic model, the dependent variable Y is regressed on the individual-level explanatory variables, as in Eq. (6.12). This estimates the values of the within-group partial slopes d_{1k} and d_{2k} and the intercepts d_{0k}. The values of Y are then adjusted by subtracting the product of each group-mean variable and its corresponding partial regression coefficient. For the two-variable illustration this would be expressed as

$$y'_{ik} = y_{ik} - d_{1k}\bar{x}_{1k} - d_{2k}\bar{x}_{2k} \tag{6.17}$$

Substituting y_{ik} from Eq. (6.17) into Eq. (6.12) we get

$$y'_{ik} + d_{1k}\bar{x}_{1k} + d_{2k}\bar{x}_{2k} = d_{0k} + d_{1k}x_{1ik} + d_{2k}x_{2ik} + f_{ik} \tag{6.18}$$

Leaving the adjusted Y variable on the left-hand side of the equation,

$$y'_{ik} = d_{0k} + d_{1k}(x_{1ik} - \bar{x}_{1k}) + d_{2k}(x_{2ik} - \bar{x}_{2k}) + f_{ik} \tag{6.19}$$

Eq. (6.19) represents the relationship between the adjusted dependent variable and the adjusted independent variables.

As in the basic model, the within-group regression coefficients are considered to be functions of group means. The difference in the centered model is that the group means are measured as deviations around the overall means for each variable. For the centered model involving two sets of explanatory variables, we have the equations

$$d_{0k} = a_{00} + a_{01}(\bar{x}_{1k} - \bar{x}_1) + a_{02}(\bar{x}_{2k} - \bar{x}_2) + u_{0k} \tag{6.20}$$

$$d_{1k} = a_{10} + a_{11}(\bar{x}_{1k} - \bar{x}_1) + a_{12}(\bar{x}_{2k} - \bar{x}_2) + u_{1k} \tag{6.21}$$

$$d_{2k} = a_{20} + a_{21}(\bar{x}_{1k} - \bar{x}_1) + a_{22}(\bar{x}_{2k} - \bar{x}_2) + u_{2k} \tag{6.22}$$

As before, the slopes in the centered model are not affected by the procedure, but the intercepts are.

The single equation is obtained by substituting the d's from Eqs. (6.20), (6.21), and (6.22) into Eq. (6.19):

$$\begin{aligned} y'_{ik} = {} & a_{00} + a_{10}(x_{1ik} - \bar{x}_{1k}) + a_{20}(x_{2ik} - \bar{x}_{2k}) \\ & + a_{01}(\bar{x}_{1k} - \bar{x}_1) + a_{02}(\bar{x}_{2k} - \bar{x}_2) \\ & + a_{11}(x_{1ik} - \bar{x}_{1k})(\bar{x}_{1k} - \bar{x}_1) + a_{12}(x_{1ik} - \bar{x}_{1k})(\bar{x}_{2k} - \bar{x}_2) \\ & + a_{21}(x_{2ik} - \bar{x}_{2k})(\bar{x}_{1k} - \bar{x}_1) + a_{22}(x_{2ik} - \bar{x}_{2k})(\bar{x}_{2k} - \bar{x}_2) \\ & + e'_{ik} \end{aligned} \tag{6.23}$$

where

$$e'_{ik} = f_{ik} + u_{0k} + u_{1k}(x_{1ik} - \bar{x}_{1k}) + u_{2k}(x_{2ik} - \bar{x}_{2k}) \tag{6.24}$$

6.2.5. Evaluating Effects in the Case of More Than One Set of Explanatory Variables with the Centered Model

Centering reduces or eliminates correlations between explanatory variables, thereby enabling less ambiguous statements about the relative contributions of the variables. It is assumed here that within-group variances of the individual-level explanatory variables do not vary from group to group. A discussion of general conditions is contained in Appendix G. After centering the model with more than one set of explanatory variables, all variables located at one level (individual, group, or interaction) will be uncorrelated with all variables located at the other levels. The centering operations *do not* remove correlations between variables on the *same* levels. While the analysis cannot be summarized in terms of the unique contributions of each variable separately, the analysis can be summarized in terms of the unique contributions of each of the three *levels* of variables. This idea can be expressed in an equation by combining variables on the same level using summation signs. Thus,

$$\begin{aligned} y'_{ik} &= a_{00} + \sum_{v} a_{v0}(x_{vik} - \bar{x}_{vk}) + \sum_{v} a_{0v}(\bar{x}_{vk} - \bar{x}_{v}) \\ &+ \sum_{r}\sum_{s} a_{rs}(x_{rik} - \bar{x}_{rk})(\bar{x}_{sk} - \bar{x}_{s}) \\ &+ f_{ik} + u_{0k} + \sum_{v} u_{vk}(x_{vik} - \bar{x}_{vk}) \end{aligned} \tag{6.25}$$

In Eq. (6.25), all individual-level variables are contained in the first sum, the group variables are contained in the second sum, and the third sum contains all the interaction variables. There are V terms for the individual variables, where V represents the number of individual-level explanatory variables. There are also V terms for the group variables, and there are V^2 terms for the cross-level interaction variables. In the preceding example there were two group-level variables and four interaction variables. If the model also contained terms for higher degree group or interaction effects, as described earlier in this chapter, they would be included in the appropriate group or interaction set.

This framework leads to the sum-of-squares table (Table 6.5) which contains abbreviations for the sums of squares described here. Starting with the corner of the table, the total sum of squares (TSS) indicates the total variation of the dependent variable. This is obtained from

$$\text{TSS} = \sum_{i}\sum_{k} (y'_{ik} - \bar{y}')^2 \tag{6.26}$$

For the total explained sum of squares,

$$\text{TSS} - \text{TRSS} = \sum_{i}\sum_{k} (\hat{y}'_{ik} - \bar{y}')^2 \tag{6.27}$$

This measure of the total variation explained by all the variables together is obtained from Eq. (6.23).

Table 6.5
Sum-of-squares Table for the Case Involving More Than One Set of Explanatory Variables

	Explained	Unexplained	Total
Individual variables	IndSS	IndRSS	IndSS + IndRSS
Group variables	GSS	GRSS	GSS + GRSS
Interaction variables	IntSS	IntRSS	IntSS + IntRSS
Total	TSS − TRSS	TRSS	TSS

Because the centering procedure makes the three levels of variables uncorrelated, the explained sums of squares uniquely associated with each level can be obtained by regressing Y' on each set of variables separately.

As before, the second column of Table 6.5 partitions the variation left unexplained by the variables specified in the model. After centering, the residuals associated with the within-group regression f_{ik} will be uncorrelated with the other sets. For that part of the total variation traceable to unspecified individual-level variables,

$$\text{IndRSS} = \sum_i \sum_k f_{ik}^2 \qquad (6.28)$$

Residuals associated with the group level u_{0k} will also be uncorrelated. For that part of the unexplained sums of squares,

$$\text{GRSS} = \sum_k n_k u_{0k}^2 \qquad (6.29)$$

The residual sums of squares associated with interactions can now be obtained by subtracting the individual- and group-level residual sums (IndRSS + GRSS) from the total residual sums (TRSS).

As in the centered basic model, the third column of the analysis-of-variance table is obtained by adding the first two columns across the rows.

6.3
Summary

The basic model of contextual analysis was generalized to the case where context effects are not linear in form. This is developed using a polynomial of the second degree in the equations involving the within-group intercepts and slopes. This extension is often important substantively in order to understand how the context is involved in a particular study. A special point is that the presence of nonlinear context effects is the primary reason for getting different estimates from the separate- and single-equation procedures.

This chapter also shows that the basic model of contextual analysis can be generalized to analyses involving more than one set of explanatory variables. Multicollinearity complicates the evaluation of results in terms of the relative importance of the explanatory variables and, therefore, requires a stepwise analysis of variation where the variables are introduced in blocks. When the centering procedures are generalized to this case, the correlations between levels of variables are removed. The analysis can then be summarized in terms of three unique parts, each associated with one of the three levels.

Contextual Analysis with More Than One Group Level of Observation

The basic model of contextual analysis, with its extensions to nonlinear context effects and more than one set of explanatory variables, explores the effects of belonging to *one* group. But individuals generally belong to more than one group and are subject to multiple, often contradictory, influences.[1] This chapter extends the basic model of contextual analysis to study effects in two different multiple-group-membership situations. One occurs when groups are nested, with one group level fitting into a higher level. A student is in a classroom, several rooms make up a grade, several grades a school, several schools belong to a school district, and so forth. Nonnested or overlapping memberships occur when the individual belongs to groups whose boundaries overlap. Peer groups, neighborhoods, and school districts are possible examples of nonnested groups.

7.1 The Case Involving Nested Groups

7.1.1. Specifying the Model

Formal notation gets cumbersome for the general case for nested groups. Therefore consider a situation in which the relationship between political party identification and the choice between Republican and Democratic candidates for state governor is being studied in terms of two hierarchical sources of influence: the precinct and the state. These groups are nested because precincts do not cut across state lines. The letters *i*, *p*, and *s*, referring to individuals, precincts, and states, are

[1] Early contextual analyses were criticized on the grounds that they oversimplified reality by considering only one group type or level. This criticism along with other issues in contextual analysis are discussed in Chapter 13.

used as subscripts to illustrate the nested model. This procedure generalizes directly to cases involving more than two group levels. For example, the county could be included as another level of influence.

The broad question is whether voters of both party persuasions are influenced most in their choice of governor by local party composition or by state party composition. The more explicit objective is to determine the extent to which Republican and Democratic identifiers are influenced differently by party splits at each level of observation. What, for instance, are the effects of partisan composition on Democrats who live in precincts dominated by the Republican party, which in turn are located in a Democratic state?

The individual-level variables consist of two categories: Republican and Democratic party identification, and choice between Republican and Democratic candidates for governor. Dummy variables are defined such that the categories involving Republicans equal 1 and Democrats equal 0. The ith individual in the pth precinct in the sth state has the corresponding y_{ips} value on the dependent variable Y and the x_{ips} value on the individual-level explanatory variable X. The proportions of Republicans at each level are equal to the corresponding means of the dummy variable x_{ips}. For the sake of generalizing the continuous case, symbols for the means rather than proportions are used in this extension of the basic model of contextual analysis.

With one individual-level explanatory variable, the single equation from the basic model for the within-group relationships remains the same, except for subscript changes to identify the observation level. Observations in a particular precinct are related according to the equation

$$y_{ips} = d_{0ps} + d_{1ps}x_{ips} + f_{ips} \qquad (7.1)$$

where the d's are the regression coefficients in that precinct.

As in the basic model, the model for nested groups specifies that the slopes and intercepts in Eq. (7.1) are functions of group means plus residuals. The difference in this case is that instead of one mean for one group level there are two. In the example, the slopes and intercepts are functions of the proportions of Republicans (dummy variable means) in the precinct and the state. This is specified in the equations

$$d_{0ps} = b_{00} + b_{0p}\bar{x}_{ps} + b_{0s}\bar{x}_s + u_{ps} \qquad (7.2)$$

$$d_{1ps} = b_{10} + b_{1p}\bar{x}_{ps} + b_{0s}x_{ips}\bar{x}_s + v_{ps} \qquad (7.3)$$

Substituting the d's from Eqs. (7.2) and (7.3) into Eq. (7.1) produces the single equation

$$\begin{aligned} y_{ips} = {} & b_{00} + b_{0p}\bar{x}_{ps} + b_{0s}\bar{x}_s + b_{10}x_{ips} \\ & + b_{1p}x_{ips}\bar{x}_{ps} + b_{1s}x_{ips}\bar{x}_s \\ & + f_{ips} + u_{ps} + v_{ps}x_{ips} \end{aligned} \qquad (7.4)$$

Nonzero coefficients in Eq. (7.4) indicate the existence of effects from the individual level, the two group levels, and the interactions between the individual and

Table 7.1
Hypothetical Coefficients for a Nested-group Analysis Involving Individuals, Precincts, and States

CONSTANT b_{00}	INDIVIDUAL EFFECT b_{10}	PRECINCT EFFECT b_{0p}	STATE EFFECT b_{0s}	INTERACTION, INDIVIDUAL AND PRECINCT b_{1p}	INTERACTION, INDIVIDUAL AND STATE b_{1s}
0.00	0.45	0.00	0.20	0.35	0.00

group levels. Table 7.1 contains a set of hypothetical numerical estimates to illustrate the interpretation of results. The nonzero value of b_{10} indicates that personal party persuasion affects the choice of candidates independent of residence. Specifically, voters tend to choose a candidate whose party label corresponds to their own party identification, even after the precinct- and state-level effects of party strength are taken into consideration. Because Y and X are dummy variables, the value of 0.45 for b_{10} in Eq. (7.4) indicates that the probability of choosing a Republican gubernatorial candidate for a Republican identifier in an all Democratic precinct and an all Democratic state is still 0.45.

The zero value of b_{0p} indicates that voters are not influenced by the party split within precincts independent of personal party identification. This means that the probability of a Democrat choosing a Republican governor is not affected by party strength in precincts.

The nonzero value for b_{0s} indicates that voters are influenced by the party split at the state level. The numerical value indicates that a Democrat has a 0.20 higher probability of choosing a Republican candidate in an all Republican state than in an all Democratic state.

The nonzero value for b_{1p} reveals the existence of a cross-level effect involving personal party identification and party composition at the precinct level. This means that voters are even more likely to choose the candidate closest to their party identification when their parties are strong at the precinct level. In contrast, the zero estimate of b_{1s} indicates that the relationship between personal party identification and choice of a governor is not affected by the relative strength of political parties at the state level.

7.1.2. The Possibility of Interactions between Observation Levels

The existence of more than one group level of observation raises the possibility of effects due to particular combinations of value on the group variables. It could be theorized that voters are most influenced by party composition when parties are strong at precinct *and* state levels. This eventuality is provided for by introducing an interaction variable made of the product of the precinct and state means. The resulting model becomes

$$d_{0ps} = b_{00} + b_{0p}\bar{X}_{ps} + b_{0s}\bar{X}_s + b_{0ps}\bar{X}_{ps}\bar{X}_s + u_{ps} \tag{7.5}$$

$$d_{1ps} = b_{10} + b_{1p}\bar{X}_{ps} + b_{1s}\bar{X}_s + b_{1ps}\bar{X}_{ps}\bar{X}_s + v_{ps} \tag{7.6}$$

As before, the single equation for this specification of within-group relationships is obtained by substituting Eqs. (7.5) and (7.6) into Eq. (7.1):

$$\begin{aligned} y_{ips} = {} & b_{00} + b_{10}x_{ips} \\ & + b_{0p}\bar{x}_{ps} + b_{0s}\bar{x}_s \\ & + b_{1p}\bar{x}_{ps}x_{ips} + b_{1s}\bar{x}_s x_{ips} \\ & + b_{0ps}\bar{x}_{ps}\bar{x}_s + b_{1ps}\bar{x}_{ps}\bar{x}_s x_{ips} \\ & + f_{ips} + u_{ps} + v_{ps}x_{ips} \end{aligned} \tag{7.7}$$

Eq. (7.7) is arranged to sort out the various effects. The first line contains the coefficient for the individual effect of party identification. The second line contains the coefficients for group effects associated with precinct and state residence. The third line contains the coefficients for the cross-level interaction effects involving personal party identification and the two groups. The fourth line contains a coefficient for the interaction between the two group levels of observation and a coefficient for the second-order interaction involving individuals, precincts, and states. The last line contains the residual terms corresponding to unspecified individual, group, and interaction variables.

7.1.3. Evaluating the Relative Importance of Effects in the Model for Nested Groups

The above operations facilitate statements about the existence of effects at the specified levels, but the problem of assessing their relative importance remains. In the study of voting behavior, the relative impact of individual-, precinct-, and state-level effects is of considerable interest.

As in the model involving more than one set of explanatory variables (Section 6.2), one way to proceed is with a stepwise consideration of blocks of variables. In this instance, one block consists of the individual-level variable. Another consists of group-level variables, and the third consists of interaction variables. After deciding the order in which to introduce the blocks in the final equation, a sums-of-squares table analogous to Table 6.5 for the case involving more than one set of explanatory variables can be constructed and summarized. With multicollinearity in the explanatory variables, conclusions depend on the order in which the blocks are considered. Other avenues must be explored before firm statements about the unique contributions of various levels can be made.

Instead of introducing an interaction variable in the separate equations to determine whether the group-level variables in combination affect the dependent variable, it is possible to use the difference between successive group-level means as the context variable. For the case of precincts and states, this involves the difference between precinct means and state means, and differences between the state means and the overall (national) mean. The within-district slopes and intercepts relating personal party identification and choice of candidates are specified as functions of the deviations about the successive means as in

$$d_{0ps} = b_{00} + b_{0p}(\bar{x}_{ps} - \bar{x}_s) + b_{0s}(\bar{x}_s - \bar{x}) + u_{ps} \tag{7.8}$$

$$d_{1ps} = b_{10} + b_{1p}(\bar{x}_{ps} - \bar{x}_s) + b_{1s}(\bar{x}_s - \bar{x}) + v_{ps} \tag{7.9}$$

Substituting Eqs. (7.9) and (7.8) into Eq. (7.1) produces the single equation for this approach to the case for nested groups.

$$\begin{aligned} y_{ips} = {} & b_{00} + b_{0p}(\bar{x}_{ps} - \bar{x}_s) + b_{0s}(\bar{x}_s - \bar{x}) \\ & + b_{10}x_{ips} + b_{1p}(\bar{x}_{ps} - \bar{x}_s)x_{ips} + b_{1s}(\bar{x}_s - \bar{x})x_{ips} \\ & + f_{ips} + u_{ps} + v_{ps}x_{ips} \end{aligned} \tag{7.10}$$

The important consequence of this specification of the case for nonnested groups is that the group-level variables will be uncorrelated with each other. The explained variation therefore can be partitioned between the group levels. However, correlations between individual, group, and interaction variables remain. This problem can be resolved using the centering procedures described in Chapter 4 and the discussion in the next section.

7.1.4. The Centering Procedure for Nested Groups

Much of the centering operations in this case has already been accomplished by defining the group variables in terms of differences between the means at successively higher levels. This is done in Eqs. (7.8) and (7.9). To indicate that we are dealing with a centered model as described earlier, the *b*'s in those equations are replaced by *a*'s. Thus,

$$d_{0ps} = a_{00} + a_{0p}(\bar{x}_{ps} - \bar{x}_s) + a_{0s}(\bar{x}_s - \bar{x}) + u_{ps} \tag{7.11}$$

$$d_{1ps} = a_{10} + a_{1p}(\bar{x}_{ps} - \bar{x}_s) + a_{1s}(\bar{x}_s - \bar{x}) + v_{ps} \tag{7.12}$$

What remains is to adjust the individual-level variables as specified by the centering procedure. For the adjusted dependent variable,

$$y'_{ips} = y_{ips} - d_{1ps}x_{ips} \tag{7.13}$$

For the relationships between party identification and candidate choice within precincts,

$$y'_{ips} = d_{0ps} + d_{1ps}(x_{ips} - \bar{x}_{ps}) + f_{ips} \tag{7.14}$$

Substituting the d's from Eqs. (7.11) and (7.12) into Eq. (7.14), the single equation for the centered model becomes

$$\begin{aligned} y'_{ips} &= a_{00} + a_{0p}(\bar{x}_{ps} - \bar{x}_s) + a_{0s}(\bar{x}_s - \bar{x}) \\ &+ a_{10}(x_{ips} - \bar{x}_{ps}) + a_{1p}(\bar{x}_{ps} - \bar{x}_s)(x_{ips} - \bar{x}_{ps}) \\ &+ a_{1s}(\bar{x}_s - \bar{x})(x_{ips} - \bar{x}_{ps}) + f_{ips} + u_{ps} + v_{ps}(x_{ips} - \bar{x}_{ps}) \end{aligned} \tag{7.15}$$

The a coefficients in Eq. (7.15) have the same interpretation as the b coefficients from the uncentered model. The important difference is that all the explanatory variables including group levels of observation will be uncorrelated, provided that within-group variances do not vary substantially. This enables a unique partitioning of the explained and unexplained variation in the dependent variable.

7.1.5. The Sum-of-squares Table for Nested Groups

The interpretation and summary of effects in the nested-group model is a special case of the procedures described in Section 6.2 for the case involving more than one set of explanatory variables. When the within-group variances of the explanatory variable do not vary substantially, correlations between the explanatory variables will be removed or greatly reduced. The explained variation in Y' consequently can be divided into unique parts associated with party identification, party composition at the precinct level, party composition at the state level, and two cross-level interaction variables involving the individual- and group-level variables. The explained sums of squares are obtained by regressing the dependent variable on each explanatory variable separately. The explained sums of squares corresponding to each variable appear in the first column of the sum-of-squares table for the case of nested groups.

With the centering procedure it is also possible to partition the *unexplained variation* into unique parts corresponding to the individual level, the group levels together, and the possible cross-level interactions together. That part of the unexplained variation due to unspecified individual-level variables is found from the residuals given in Eq. (7.14) for the within-precinct relationships between party allegiance and choice for a governor. This is summarized by the sums of squares in

$$\text{IndRSS} = \sum\sum\sum f_{ips}^2 \tag{7.16}$$

where the first sum is over the number of states, the second sum is over the number of precincts, and the third sum is over the observations within districts.

The unexplained part of the variation in the dependent variable, due to unspecified group-level variables at the precinct and state levels together, is found from the residuals produced in Eq. (7.12). This group residual sum of squares is

$$\text{GRSS} = \sum\sum n_{ps} u_{ps}^2 \tag{7.17}$$

where n_{ps} is the number of observations in the pth precinct in the sth state.

The unexplained part of the variation in the dependent variable due to unspecified interaction variables is found from the residuals involving v_{ps} as they appear in Eq. (7.15). The corresponding sum of squares is obtained from

$$\text{IntRSS} = \sum\sum\sum v_{ps}^2(x_{ips} - \bar{x}_{ps})^2 \tag{7.18}$$

The centering procedure makes the three sets of residuals uncorrelated. Therefore they should add up to the total residual sum of squares obtained from the single multiple regression in Eq. (7.15). This sum is calculated from

$$\text{TRSS} = \sum\sum\sum e_{ips}'^2 = \sum\sum\sum(y_{ips}' - \hat{y}_{ips}')^2 \tag{7.19}$$

Adding the sums for both the explained and the unexplained parts of the variation at each of the three levels, we obtain the equivalent of column 3 in Table 6.5. The margins of the sum-of-squares table thus constructed for the model involving nested groups add up to the total sums of squares for the dependent variable Y', obtained from

$$\text{TSS} = \sum\sum\sum(y_{ips}' - \bar{y}')^2 \tag{7.20}$$

The third column represents the partitioning of both explained and unexplained variation into parts for the individual, precinct, state, and interaction levels.

Given the sum-of-squares table, it is possible to summarize the analysis by calculating proportions of variation looking down the columns, across the rows, and to the corner.

In many situations the within-group variances of X vary substantially. Under these circumstances, the following procedure can be used. First the sums of squares for each group-level variable are obtained. This is a good starting point, since they will be uncorrelated with each other and with the other variables in the model, regardless of whether the variances are constant. These sums can be obtained by regressing Y' on each of the group variables separately, or by using the separate equation estimates and the formulas

$$\text{precinct SS} = \sum\sum n_{ps}a_{0p}^2(\bar{x}_{ps} - \bar{x}_s)^2 \tag{7.21}$$

$$\text{state SS} = \sum n_s a_{0s}^2(\bar{x}_s - \bar{x})^2 \tag{7.22}$$

Together these sums are equivalent to GSS in Table 6.5.

The explained sum of squares for the individual-level variable is obtained by regressing Y' on the individual-level variable or from the formula

$$\text{IndSS} = a_{10}^2\sum\sum\sum(x_{ips} - \bar{x}_{ps})^2 \tag{7.23}$$

The total explained sum of squares is then obtained from Eq. (7.15), and the explained interaction sum of squares is found by subtraction. Thus,

$$\text{TregSS} = \sum\sum\sum(\hat{y}_{ips}' - \bar{y}')^2 \tag{7.24}$$

$$\text{IntSS} = \text{TregSS} - \text{GSS} - \text{IndSS} \tag{7.25}$$

When within-group variances are not constant, the partitioning of unexplained variation remains as described above.

7.2
Contextual Analysis for the Case of Overlapping Groups

With nested groups, any particular group at one level of observation is wholly contained by a group at the next higher level. All members of the smaller group can only be members of one group at each successive level. But it is possible to consider more complex stratification schemes where memberships intersect or overlap. Perhaps members of one group of one type belong to more than one group of another type. Consider a situation involving neighborhoods and school districts. It is conceivable that the residents of a particular neighborhood all belong to the same school district. This would be consistent with the nested situation, but only in this special case. It is also possible that the boundaries of a school district cut across a particular neighborhood, meaning that some residents of the *same* neighborhood will belong to one school district while others belong to another. This more complex situation is reflected in Figure 7.1.

Suppose now that we wish to consider the group effects of income levels on parental participation in school activities. The existence of overlapping group memberships raises the question of how the group-mean variables ought to be conceptualized and measured. Will an individual in subgroup 1 be most influenced by the income level of the whole neighborhood and the whole school district, or only by that portion of each where the two intersect?

Figure 7.1
An Illustration of Nonnested Groups Showing Three School Districts, Two Neighborhoods, and Four Subgroups

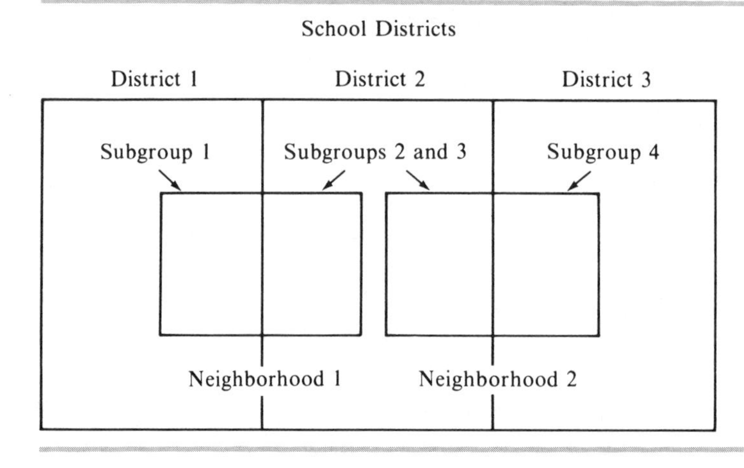

The answer depends on one's theory of how the context effects are being mediated. If the influence of the media and formal communication is emphasized, then one might construct the group means from the total membership of all groups to which an individual belongs. If, on the other hand, informal communication and interpersonal forms of influence are considered primary, then one might construct the group-mean variable on the basis of the membership in the subgroups resulting from the intersections of two or more whole groups. Implicit in this latter approach is the idea that individuals are influenced by selective parts of the groups to which they belong.

To show how both approaches can be specified as extensions of the basic model of contextual analysis, the stratification scheme for this particular example is represented in Table 7.2. It shows an area consisting of C neighborhoods and D districts for a total of $C \times D$ subgroups resulting from the intersection of neighborhoods and districts. Each subgroup has a mean of the explanatory variable X, which in this example would be the average income for the individuals in the group. To refer to these subgroups we have added the subscript j.

For this example j represents the school district, while k designates the neighborhood. Thus we have $\bar{x}_{11}, \bar{x}_{12}, \ldots$, representing the average income for the subgroups made up of the intersection of neighborhoods and school districts. The margins have the average incomes for the whole neighborhoods and school districts. These means are designated $\bar{x}_{.1}, \bar{x}_{1.}$, and so forth.

For the relationship between personal income and participation, $B \times C$ equations represent each subgroup in Table 7.2. This is represented in the equation

$$y_{ijk} = d_{0jk} + d_{1jk}x_{ijk} + f_{ijk} \qquad (7.26)$$

There has only been a change in subscripts to identify the individual's membership in two groups. Again, according to the specifications of the basic model of contextual

Table 7.2
Intersection of Neighborhoods and School Districts Showing the Means for the Explanatory Variable X

		School Districts D					
		1	2	3	...	D	
	1	$\bar{x}_{11}$	$\bar{x}_{12}$	$\bar{x}_{13}$	...	$\bar{x}_{1D}$	$\bar{x}_{1\cdot}$
	2	$\bar{x}_{21}$	$\bar{x}_{22}$	$\bar{x}_{23}$	...	$\bar{x}_{2D}$	$\bar{x}_{2\cdot}$
Neighborhoods C	3	$\bar{x}_{31}$	$\bar{x}_{32}$	$\bar{x}_{33}$	...	$\bar{x}_{3D}$	$\bar{x}_{3}$
	⋮	⋮	⋮	⋮		⋮	⋮
	C	$\bar{x}_{C1}$	$\bar{x}_{C2}$	$\bar{x}_{C3}$	...	$\bar{x}_{CD}$	$\bar{x}_{\cdot C}$
		$\bar{x}_{\cdot 1}$	$\bar{x}_{\cdot 2}$	$\bar{x}_{\cdot 3}$	...	$\bar{x}_{\cdot D}$	$\bar{x}$

analysis, we expect the d's in Eq. (7.26) to be functions of group-mean variables. The difference from the simple case and the case for nested groups is that a decision has to be made regarding the construction of the group-mean variables. One way is to treat the d's as functions of the means constructed in the margins of Table 7.2. For the example this suggests a study of the influence of the average income for the whole school district *and* the whole neighborhood on individual-level participation rates. This suggests the separate equations

$$d_{0jk} = b_{00} + b_{01}\bar{x}_{j\cdot} + b_{02}\bar{x}_{\cdot k} + u_{0jk} \tag{7.27}$$

$$d_{1jk} = b_{10} + b_{11}\bar{x}_{j\cdot} + b_{12}\bar{x}_{\cdot k} + u_{1jk} \tag{7.28}$$

The single equation for the analysis is obtained as before by substituting the d's from Eqs. (7.27) and (7.28) into Eq. (7.26), giving

$$\begin{aligned} y_{ijk} = {} & b_{00} + b_{01}\bar{x}_{j\cdot} + b_{02}\bar{x}_{\cdot k} + b_{10}x_{ijk} + b_{11}x_{ijk}\bar{x}_{j\cdot} \\ & + b_{12}x_{ijk}\bar{x}_{\cdot k} + f_{ijk} + u_{0jk} + u_{1jk}x_{ijk} \end{aligned} \tag{7.29}$$

From Eq. (7.29) we can determine the existence of individual-level effects, group effects originating in the neighborhoods, group effects originating in the school districts, and cross-level interactions involving the individual and group variables. It is conceivable that individuals residing in high-income neighborhoods *and* high-income school districts (or vice versa) experience stronger influences from the environment than if their neighborhoods differed significantly from their school districts with respect to income. This eventuality is addressed by defining an interaction term made of the product of the two group means $\bar{x}_{j\cdot}$ and $\bar{x}_{\cdot k}$.

Another way to conceptualize the case of overlapping groups is to use as group means the average incomes of the subgroups based on the intersections of neighborhoods and school districts. Substantively, perhaps individuals receive their signals from a more selective environment of just those individuals residing in the same subgroup. This is formally represented in the equations

$$d_{0jk} = b_{00} + b_{01}\bar{x}_{jk} + u_{0jk} \tag{7.30}$$

$$d_{1jk} = b_{10} + b_{11}\bar{x}_{jk} + u_{1jk} \tag{7.31}$$

To obtain the single equation, the d's from Eqs. (7.30) and (7.31) are substituted into Eq. (7.26).

7.2.1. Evaluating the Relative Importance of Effects in the Case of Overlapping Groups

As in previous cases, it is possible to apply a stepwise procedure to the partitioning of explained and unexplained variation between the levels specified in the analysis. The individual variable may be introduced first, one of the (overlapping) group variables second, the other group variable third, and potential interactions last.

The problem is that the order in which variables are considered in the final model frequently determines conclusions about their relative importance. This is because of multicollinearity in the explanatory variables. There is again a need to extend the centering procedures described for the basic model.

Generally, the centering procedures for the case of overlapping groups is a special case of the centering procedures for more than one set of explanatory variables (cf. Section 6.2.4). The difference is that there are one individual variable and two (or more) group and interaction variables. Procedures for constructing and interpreting the sum-of-squares table for nonnested groups are identical. To illustrate, the values of the individual-level variables can be adjusted as prescribed by the centering procedure, and their relationships within groups can be expressed in the equation

$$y'_{ijk} = d_{0jk} + d_{1jk}(x_{ijk} - \bar{x}_{jk}) + f_{ijk} \tag{7.32}$$

The effect of the neighborhood with respect to income levels is measured by the deviation of the neighborhood means from the overall mean, that is, we have a new variable $(\bar{x}_{j\cdot} - \bar{x})$. The effect of the school district is measured in the same way, through $(\bar{x}_{\cdot k} - \bar{x})$. The third group variable, based on the subgroup means, is based on the usual interaction term in two-way analysis of variance $(\bar{x}_{jk} - \bar{x}_{j\cdot} - \bar{x}_{\cdot k} + \bar{x})$. From the general logic of the contextual model we get the separate equations

$$d_{0jk} = a_0 + a_2(\bar{x}_{j\cdot} - \bar{x}) + a_4(\bar{x}_{\cdot k} - \bar{x}) + a_6(\bar{x}_{jk} - \bar{x}_{j\cdot} - \bar{x}_{\cdot k} + \bar{x}) + u_{jk} \tag{7.33}$$

$$d_{1jk} = a_1 + a_3(\bar{x}_{j\cdot} - \bar{x}) + a_5(\bar{x}_{\cdot k} - \bar{x}) + a_7(\bar{x}_{jk} - \bar{x}_{j\cdot} - \bar{x}_{\cdot k} + \bar{x}) + v_{jk} \tag{7.34}$$

From Eqs. (7.32), (7.33), and (7.34) we get the single equation showing how the centered variable Y' relates to the various explanatory variables constructed above. Thus,

$$\begin{aligned} y'_{ijk} = a_0 &+ a_1(x_{ijk} - \bar{x}_{jk}) \\ &+ a_2(\bar{x}_{j\cdot} - \bar{x}) + a_4(\bar{x}_{\cdot k} - \bar{x}) \\ &+ a_6(\bar{x}_{jk} - \bar{x}_{j\cdot} - \bar{x}_{\cdot k} + \bar{x}) \\ &+ a_3(x_{ijk} - \bar{x}_{jk})(\bar{x}_{j\cdot} - \bar{x}) \\ &+ a_5(x_{ijk} - \bar{x}_{jk})(\bar{x}_{\cdot k} - \bar{x}) \\ &+ a_7(x_{ijk} - \bar{x}_{jk})(\bar{x}_{jk} - x_{j\cdot} - \bar{x}_{\cdot k} + \bar{x}) \\ &+ f_{ijk} + u_{jk} + v_{jk}(x_{ijk} - \bar{x}_{jk}) \end{aligned} \tag{7.35}$$

Eq. (7.35) shows that Y' is influenced by one individual-level variable, three group-level variables of X resulting from the two stratification variables, and three individual–group interaction variables.

7.2.2. The Sum-of-squares Table for Overlapping Groups

The results of this analysis can be summarized by constructing a sum-of-squares table of the type used in each of the preceding model extensions. Each variable located at one of the three levels of observation, that is, individual, group, and interaction, is uncorrelated with each variable at another level when the within-group variances of X are equal. But in this case variables located at the same level cannot be expected to be uncorrelated with each other, except in the special case where the number of observations in each group are identical or at least proportional. Nevertheless, the analysis can proceed by getting the sum of squares associated with the group level from

$$\text{GSS} = \sum\sum n_{jk}[a_2(\bar{x}_{j\cdot} - \bar{x}) + a_4(\bar{x}_{\cdot k} - \bar{x}) + a_6(\bar{x}_{jk} - \bar{x}_{j\cdot} - \bar{x}_{\cdot k} + \bar{x})]^2 \quad (7.36)$$

The sum of squares for the individual level is obtained as before from

$$\text{IndSS} = a_1^2\sum\sum\sum(x_{ijk} - \bar{x}_{jk})^2 \quad (7.37)$$

The sum of squares associated with the interaction level is obtained by subtracting the group- and the individual-level sums of squares from the total regression sum of squares. The rest of the table can then be constructed just as in the cases previously described for the other model generalizations.

7.3 Summary

The basic model of contextual analysis described in Chapter 3 considers the effects of belonging to just one group. This chapter shows how it is possible to extend that model to study the effects of belonging to more than one group. Two multiple-group-membership situations are considered. One involves groups that are nested with one group level fitting into a higher level. Precincts and states are used to illustrate concepts and procedures. It is shown how effects on a dependent individual variable could be traced to the various levels. The second multiple-group-membership situation involves the case where individuals belong to groups whose boundaries overlap. An example concerning neighborhoods and school districts illustrates this extension of the basic model.

III

Explaining Aggregate Relationships

Part II deals with individual relationships in terms of individual and group variables. In Part III we study aggregate-level relationships in terms of individual-level behavior. The multilevel analysis objective of explaining aggregate relationships in terms of individual behavior is introduced in Chapter 2. Chapter 8 elaborates the concepts and procedures described there. This includes the development of a formal framework and extensions to more complex situations. Chapter 9 presents additional substantive examples of these concepts and procedures. Suggestions for computer assistance are contained in Appendix A.

Explaining Aggregate Relationships in Terms of Individual Behavior

This chapter examines concepts and procedures for explaining aggregate relationships.[1] Two cases are considered. The first involves explaining a relationship between group-mean variables. The second involves explaining a relationship observed in a population of individuals across several groups. A formal model of multilevel structure is presented which contains parameters for aggregate relationships and for individual and group effects.

The multilevel structure of relationships demonstrates how the signs and numerical values of the group and population slopes depend on the specific signs and magnitudes of individual and group effects without interaction and on the ratio of variations for the individual and group variables. This is followed by examining the more complex cases involving a cross-level interaction effect and nonlinear context effects. Computer assistance for this type of multilevel analysis involves combinations of the procedures described in Appendix A.

8.1 The Consequences of Incorrect Model Specification on Aggregate Relationships

Consider a hypothetical study involving sales departments in five districts of a private industry. The focus of the study is the relationship between competitiveness and sales. The dependent variable Y is a measure of personal sales, and X is a measure

[1] The References at the end of this book identify many works directly or indirectly related to the problems of explaining aggregate relationships. For a list of works more directly pertinent to this chapter, see footnote 1, Chapter 2.

Table 8.1
Data for the Hypothetical Study of Sales Groups and Competitiveness

DISTRICT NUMBER k	PERSONAL SALES y_{ik}	COMPETITIVENESS SCORES x_{ik}	DISTRICT AVERAGES $\bar{y}_k$	$\bar{x}_k$	WITHIN-DISTRICT REGRESSION LINES $y_{ik} = d_{0k} + d_{1k}x_{ik} + f_{ik}$
1	55	1	65	3	$y_{i1} = 50.0 + 5.0x_{i1} + f_{i1}$
	80	3			
	50	3			
	75	5			
2	50	2	60	4	$y_{i2} = 40.0 + 5.0x_{i2} + f_{i2}$
	70	4			
	50	4			
	70	6			
3	45	3	55	5	$y_{i3} = 30.0 + 5.0x_{i3} + f_{i3}$
	65	5			
	45	5			
	65	7			
4	40	4	50	6	$y_{i4} = 20.0 + 5.0x_{i4} + f_{i4}$
	60	6			
	40	6			
	60	8			
5	35	5	45	7	$y_{i5} = 10.0 + 5.0x_{i5} + f_{i5}$
	55	7			
	35	7			
	55	9			

of competitiveness. The measure of Y is based on a scale ranging from 0 to 100. The measure of X is based on a scale ranging from 0 to 10. Low values indicate low competitiveness and low sales. Observed scores on these variables are presented in Table 8.1.

To study the relationship between competitiveness and sales, the first three approaches presented in Sections 1.1, 1.2, and 1.3 to study individual behavior are applied. The first involves regressing Y on X within each of the districts. The equation is

$$y_{ik} = d_{0k} + d_{1k}x_{ik} + f_{ik} \tag{8.1}$$

where the subscript k identifies the district number. The numerical coefficients for each district are given in Table 8.1.

The second approach is to regress Y on X for the observations in the five groups together. This is referred to as the population regression. The slope estimate is referred to as the population slope. The equation is

$$\begin{aligned} y_{ik} &= d_{0p} + d_{1p}x_{ik} + e_{ik} \\ &= 55.0 + 0.0x_{ik} + e_{ik} \end{aligned} \tag{8.2}$$

where the subscript p indicates that the regression uses the observations for Y and X in all (five) districts.

The third approach is to regress the average sales for each district on the level of competitiveness (the district averages of X). This is referred to as the group-level regression. The slope estimate is referred to as the "group-level slope." The equation is

$$\begin{aligned} \bar{y}_k &= d_{0g} + d_{1g}\bar{x}_k + e_k \\ &= 80.0 - 5.0\bar{x}_k + e_k \end{aligned} \tag{8.3}$$

where the subscript g indicates that the regression applies to groups (sales districts).

Examination of the above results immediately raises questions. As intuition suggests, the relationships between Y and X within districts are positive and moderately strong. However, we are struck by the fact that there is no relationship between Y and X at the population level. To compound the puzzle, the relationship between district sales levels and levels of competitiveness is strongly negative. This is counter to the expectation that the individual-level relationship between sales and competitiveness should average up to the population relationship and the relationship between sales districts. What accounts for this phenomenon? The answer leads to an explanation of the population and group-level relationships.

The unexpected results are obtained because the model of individual behavior implicit in our first expectations is incorrectly specified. Section 3.2.1 describes the special case of individual effects only. The specification there makes explicit what was implicit in our first expectations. That special case assumes that both the intercepts d_{0k} and the slopes d_{1k} are constant across groups (excepting only random effects). Estimates for this model with the sales district data are

$$d_{0k} = b_0 + u_k = 30.0 + u_k \tag{8.4}$$

$$d_{1k} = b_1 + v_k = 5.0 + v_k \tag{8.5}$$

The single equation for the case of individual effects

$$\begin{aligned} y_{ik} &= b_0 + b_1 x_{ik} + e_{ik} \\ &= 30.0 + 5.0x_{ik} + e_{ik} \end{aligned} \tag{8.6}$$

is of the same form as the population equation [Eq. (8.2)]. If the model of individual behavior specified in Eqs. (8.4) and (8.5) is correct, then the coefficients b_0 and b_1 should equal d_{0p} and d_{1p}. This clearly is not the case. Compare 55.0 to 30.0 for the intercept and 5.0 to 0.0 for the slope.

The implications of an incorrectly specified model of individual behavior also can be identified for the observed group-level relationship. Eq. (8.6) is aggregated to the level of districts by summing and averaging Y, X, and E within districts, producing

$$\begin{aligned} \bar{y}_k &= b_0 + b_1\bar{x}_k + \bar{e}_k \\ &= 30.0 + 5.0\bar{x}_k + \bar{e}_k \end{aligned} \tag{8.7}$$

Eq. (8.7) is of the same form as the group-relationship equation [Eq. (8.3)]. If the model of individual behavior specified in Eqs. (8.4) and (8.5) is correct, the coefficients b_0 and b_1 should equal the coefficients d_{0g} and d_{1g}. In this instance the slopes have opposite signs. Compare 30.0 to 80.0 for the intercept and 5.0 to -5.0 for the slope.

8.2
Explaining the Group-level Relationship in Terms of Individual-level Behavior

Understanding the group-level relationship reflected in d_{0g} and d_{1g} requires a correctly specified model of individual behavior. This is obtained through a contextual analysis as described in Section 1.7. That approach is distinguished by its attention to the role of the group-mean variable $\bar{X}$ in explaining individual behavior. As an indicator of competitive climates in sales districts, the group mean could have different implications for personal sales than individual competitiveness. For example, rising *levels* of competitiveness could adversely affect other factors such as cooperation, communication, and group solidarity, with negative consequences for sales.

The separate equations in this model result in

$$\begin{aligned} d_{0k} &= b_0 + b_2\bar{x}_k + u_k \\ &= 80.0 - 10.0\bar{x}_k + u_k \end{aligned} \tag{8.8}$$

$$\begin{aligned} d_{1k} &= b_1 + b_3\bar{x}_k + v_k \\ &= 5.0 + 0.0\bar{x}_k + v_k \end{aligned} \tag{8.9}$$

The single equation becomes

$$\begin{aligned} y_{ik} &= b_0 + b_1 x_{ik} + b_2\bar{x}_k + b_3 x_{ik}\bar{x}_k + e_{ik} \\ &= 80.0 + 5.0x_{ik} - 10.0\bar{x}_k + 0.0x_{ik}\bar{x}_k + e_{ik} \end{aligned} \tag{8.10}$$

The results indicate the presence of individual and group effects. The zero value for the coefficient b_3 indicates no interaction effect. This is the model described in Section 3.2.4.

The difference between the single equation for this model and Eq. (8.6) is that it contains terms for individual and group effects. Aggregating Eq. (8.10), rather than Eq. (8.6), to the level of districts, produces

$$\bar{y}_k = b_0 + b_1\bar{x}_k + b_2\bar{x}_k + \bar{e}_k \tag{8.11}$$

Since the terms b_1 and b_2 are associated with the same term $\bar{x}_k$, they are collected to give

$$\bar{y}_k = b_0 + (b_1 + b_2)\bar{x}_k + \bar{e}_k \tag{8.12}$$

Eq. (8.12) is the same form as the equation for the relationship between the group means [Eq. (8.3)]. If the model of individual behavior is correctly specified, the two sets of coefficients should be equal. Specifically,

$$\begin{aligned} d_{0g} &= b_0 = 80.0 \\ d_{1g} &= b_1 + b_2 = (5.0 - 10.0) = -5.0 \end{aligned} \tag{8.13}$$

Since this is true, it can be said that the group-level relationship has been *explained* in terms of individual behavior. Specifically, an individual effect of competitiveness operated in the expected (positive) direction. However, the group effect of competitiveness operated in the opposite direction.[2] Substantively, the negative group effect might reflect the consequences of reduced cooperation, disrupted communications, and reduced group solidarity arising with higher levels of competitiveness. The negative group effect was larger in numerical value than the positive individual effect. The net effect explains the "puzzling" negative slope observed in the group-level relationship.

8.3

Explaining the Population Relationship in Terms of Individual-level Behavior

In the preceding discussion, the group-level coefficients d_{0g} and d_{1g} were explained in terms of the sign and magnitude of individual and group effects. The problems remain to explain why there was no relationship at the population level and why the group relationship and the population relationship were different. These questions require the elaboration of a multilevel model containing terms for the population slope d_{1p}, the group-level slope d_{1g}, the measures of individual and group effects (b_1 and b_2), and the variances of X and $\bar{X}$. The relationships specified in this model are referred to as the *multilevel structure*. Given this structure, the population slope can be explained in terms of the group-level slope and effects on individual behavior.

This model is based on the contextual model in Section 3.2.4 which specified an individual effect and a group effect only. It is derived by partitioning the population slope into within- and between-group parts. This introduces the slopes d_{1p} and d_{1g}. The magnitude of the individual effect is introduced by substituting d_{1k} from Eq. (8.5) into the within-group part of the equation. The estimate of the group effect is visible in the group slope d_{1g}, which equals the sum of the individual and the group effects. The relevant parts of the model are presented here. The full derivation is contained in Appendix H.

[2] Cases where individual and group effects operated in opposite directions were observed in real data by Kendall and Lazarsfeld (1950) and Davis et al. (1961).

The model is based on the following considerations. The total variation in the explanatory variable X can be measured by the total sum of squares $\text{TSS} = \sum\sum(x_{ik} - \bar{x})^2$. The variation in the group means can be measured by the between sum of squares $\text{BSS} = \sum n_k(\bar{x}_k - \bar{x})^2$. The proportion of the total variation accounted for by the groups is denoted by η^2 (eta squared), where $\eta^2 = \text{BSS/TSS}$. The regression coefficients, η^2 and $(1 - \eta^2)$ are then related according to the equations

$$\begin{aligned} d_{1\,g} &= b_1 + b_2 \\ -5.0 &= 5.0 - 10.0 \end{aligned} \tag{8.14}$$

$$\begin{aligned} d_{1\,p} &= b_1 + b_2\eta^2 \\ 0.0 &= 5.0 - 10.0(0.5) \end{aligned} \tag{8.15}$$

$$\begin{aligned} d_{1\,p} &= b_1(1 - \eta^2) + d_{1\,g}\eta^2 \\ 0.0 &= 5.0(1 - 0.5) - 5.0(0.5) \end{aligned} \tag{8.16}$$

$$\begin{aligned} d_{1\,p} &= d_{1\,g} - b_2(1 - \eta^2) \\ 0.0 &= -5.0 - 10(1 - 0.5) \end{aligned} \tag{8.17}$$

Eq. (8.13), relating the group slope to the group and interaction effects, is repeated as Eq. (8.14) for convenience. Numerical values based on the sales district data are given with each equation. The values for b_1 and b_2 in Eqs. (8.14) through (8.17) are obtained from Eqs. (8.8) and (8.9). For the sales district data they are 5.0 and -10.0. The zero value of $d_{1\,p}$, obtained from the population regression in Eq. (8.2), needs to be explained.

Eq. (8.15) indicates that the numerical value of the population slope equals the sum of the individual effects plus a fraction of the group effect. The fraction based on η^2 equals 1 at the theoretical point where all the total variation in X is attributed to between-group differences. We have $d_{1\,p}$ equal to 0.0. Also, $b_1 + b_2\eta^2$ equals 0.0 as seen in Eq. (8.17). Thus these data satisfy Eq. (8.17) and the aggregate relationship therefore has been explained in terms of individual behavior. It follows that X and Y are related according to the equation

$$y_{ik} = d_{0\,p} + 0.0x_{ik} + e_{ik} \tag{8.18}$$

Specifically, there is a positive individual-level effect of competitiveness on sales, and there is a negative group effect of district-competitiveness levels on sales. The group effect is numerically twice as large as the individual effect. However, only half of the negative effect counteracts the positive individual effect in the generation of the population slope. This is due to homogeneity between groups relative to homogeneity within groups. In this instance the configuration of group and individual effects and the ratio of sums of squares (η^2) produces a zero population relationship.

The relationship at the group level is generally recognized as an aggregate phenomenon because individual variables are summarized and observed at higher levels of observation. Group-level relationships have been the object of considerable attention as a source of mischief when used to make inferences about individuals.[3] In contrast, the population relationship is not generally recognized as an aggregate phenomenon. That is because observations are analyzed in their unsummarized form. Also, population relationships are not generally recognized as something that requires explanation in the same way as other aggregate phenomena. The discussion in this section demonstrates that the population relationship is an aggregate phenomenon because it is the consequence of adding up individual and group effects across a number of groups. Like other aggregate relationships, it too requires explanation in terms of models of individual behavior.

8.4

Explaining How the Group-level Slope and the Population Slope Differ

Why are the aggregate slopes d_{1g} and d_{1p} different in the sales-group data? The answer is implicit in the explanations for each described above. The related questions of when the two aggregate relationships are the same and how they are related is made explicit in Eq. (8.17). Specifically, the coefficient d_{1p} equals d_{1g} when b_2 equals zero. This means that the population slope will equal the group slope when there is no group effect. Eq. (8.17) also indicates that the two aggregate slopes are equal when the term $(1 - \eta^2)$ equals zero. In this situation the variation in the group variable $\bar{X}$ accounts for all of the variation in X.

Generally η^2 will not equal one or zero, and the group effect will not be zero. Therefore the aggregate slopes will generally have different values. The model of multilevel structure in Eqs. (8.14) through (8.17) makes explicit how changes in η^2 will increase or decrease the difference in the aggregate coefficients. A comparison of Eqs. (8.14) and (8.15) shows that, when the individual and group effects have the same sign, an increase in η^2 will decrease the difference between the two aggregate coefficients.

The behavior of the two slopes, given different η^2, also can be observed directly in graphs similar to those used in Chapter 3 to represent within-group regression lines. The group variables $\bar{X}$ and $\bar{Y}$ are direct transformations of the individual variables X and Y. This makes it possible to plot the regression line for the group-level relationship on the same graph used to display the regression lines for each group and the regression line for the whole population. The convenient result is that the entire multilevel structure can be represented in a single graph.

Figure 8.1 presents the results for the sales-group study. Although residuals are not shown, the lengths of the within-group regression lines and their positions on the X axis approximate η^2. The value of η^2 is 0.50.

[3] It is generally assumed that researchers became aware of the "ecological fallacy" after Robinson (1950). However, the practice of using group data to make individual level inferences did not disappear completely.

Figure 8.1
Graph of Multilevel Structure for X and Y

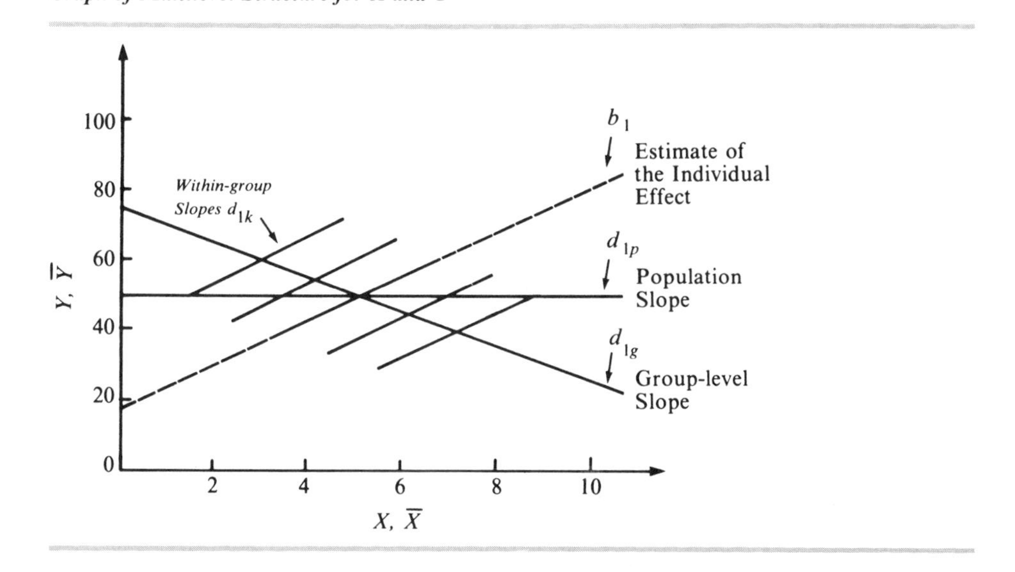

To visualize the effect of larger or smaller ratios on the relative sizes of the two aggregate slopes, two hypothetical situations are shown in Figure 8.2. Note that the lines with slopes d_{1p} and d_{1g} intersect at their common coordinates $(\bar{x}_k, \bar{y}_k)$. Because of this, the two slopes act like a pair of scissors, opening and closing with changes in η^2. As the ratio approaches unity, the value of the population slope approaches the value of the group slope. When η^2 approaches zero, the difference between the aggregate slopes approaches a maximum, which in this case is the value of the group effect b_2.

8.5

Explaining How the Population Slope Varies Relative to the Estimate of the Individual Effect

The population slope cannot generally be considered equivalent to a measure of the individual effect. However, it is reasonable to ask when it is equivalent. Eq. (8.15) indicates that the answer involves the coefficient of the group effect b_2 and η^2. The value of the population slope and the value of the individual effect will be equal when there is no group effect and, in the extreme case, where there is no between-group variation. However, the more general case will involve a group effect and a nonzero η^2.

Figure 8.2
Variations in Slopes with Changes in η^2. (a) η^2 is large. (b) η^2 is small

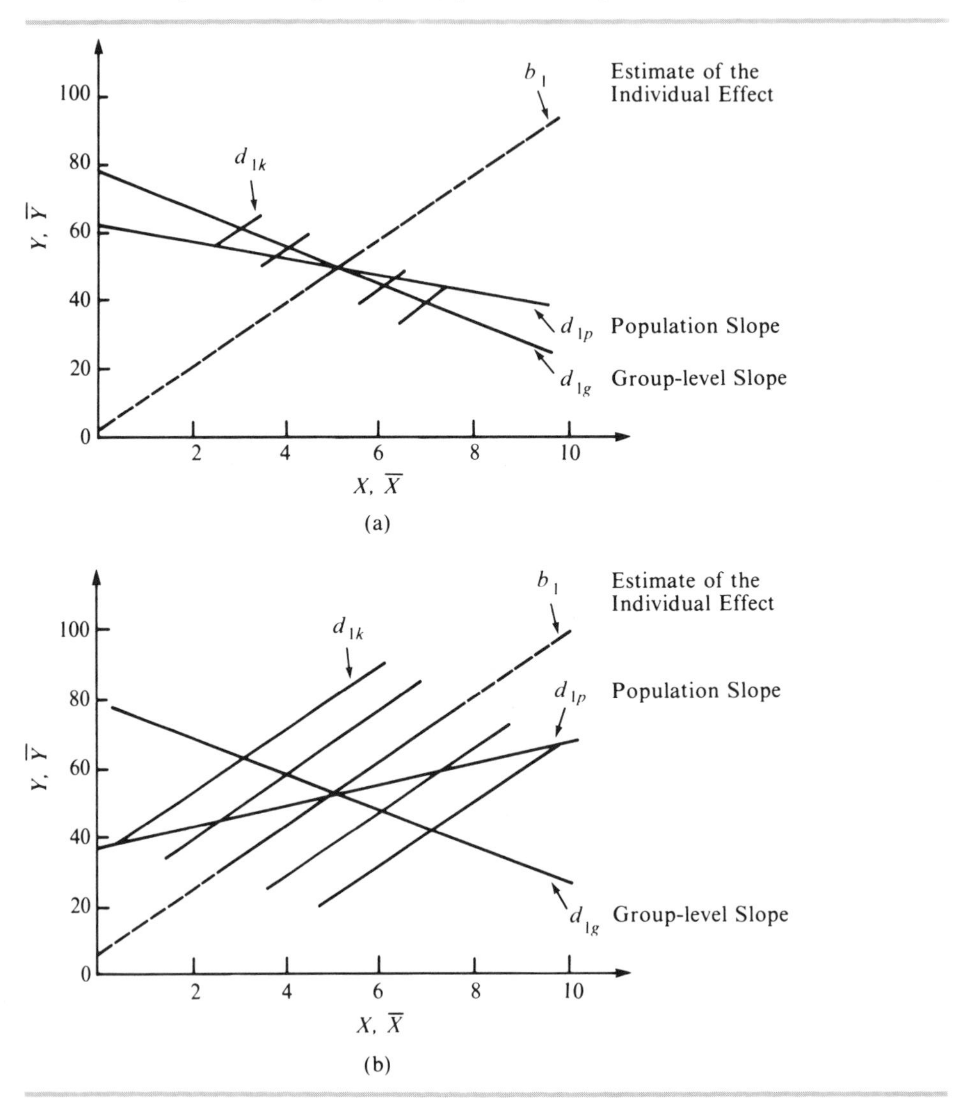

As before, the behavior of the population slope relative to the coefficient for the individual effect can be observed graphically in Figure 8.2. Under the assumption of no interaction effect, the within-group slopes d_{1k} are equal to b_1. For comparison to the other slopes, a single line is used to indicate the individual effect b_1. Compare Figure 8.2(a) and (b). Notice that the larger η^2 moves the population slope closer to the group-level slope and *away* from the coefficient for the individual effect. The illustrations in Figure 8.2 also show that the population slope always lies *between* the group-level slope and the coefficient for the individual effect. It is closer to the coefficient for the individual effect, when η^2 is small. It is closer to the group-level slope when η^2 is large.

Since the coefficient for an aggregate relationship might be used as an indicator of an individual or group effect, understanding relationships between the aggregate coefficients, coefficients for individual and group effects, and η^2 is important when making cross-level inferences. It is also important to understand the relationships when selecting the appropriate level to define group variables. Should the groups be census tracts, cities, counties, states, or regions? How will a wrong choice affect results? If choosing successively larger units as groups reduces the sum of squares of $\bar{X}$ relative to the total sum of squares, the coefficient for the individual effect b_1 will be inflated. Since the slopes d_{1g} and d_{1p} are unaffected in this case, the coefficient for the group effect b_2 will be underestimated. The effects of changing η^2 also arises when continuous group-level variables are dichotomized, as in the Blau method. In such a case the group sum of squares is artifically reduced, again underestimating the group effect.

8.6 The Implications of a Cross-level Interaction Effect in the Explanation of Aggregate Relationships

An aggregate-level relationship is explained when the aggregation of a model of individual behavior reproduces that relationship. This occurs when the model of individual behavior is correctly specified. In the preceding example the correct model contained no cross-level interaction effect. To illustrate what happens when an interaction effect is ignored, consider the data set given in Table 8.2.

Applying the individual and group effect model produces the following results. For the population intercept and slope,

$$\begin{aligned} y_{ik} &= d_{0p} + d_{1p} x_{ik} + e_{ik} \\ &= 29.2 + 6.4 x_{ik} + e_{ik} \end{aligned} \tag{8.19}$$

For the group-level intercept and slope.

$$\begin{aligned} \bar{y}_k &= d_{0g} + d_{1g} \bar{x}_k + e_k \\ &= 31.0 + 6.0 \bar{x}_k + e_k \end{aligned} \tag{8.20}$$

For the coefficients b_0, b_1, and b_2,

$$\begin{aligned} d_{0k} &= b_0 + b_2 \bar{x}_k + u_k \\ &= 50.0 - 5.0 \bar{x}_k + u_k \end{aligned} \tag{8.21}$$

$$\begin{aligned} d_{1k} &= b_1 + v_k \\ &= 6.5 + v_k \end{aligned} \tag{8.22}$$

Table 8.2
Data for the Hypothetical Study of Sales Groups Illustrating the Implications of an Interaction Effect in the Explanation of Aggregate Relationships

DISTRICT NUMBER k	PERSONAL SALES y_{ik}	COMPETITIVENESS SCORES x_{ik}	DISTRICT AVERAGES $\bar{y}_k$	$\bar{x}_k$	WITHIN-DISTRICT REGRESSION LINES $y_{ik} = d_{0k} + d_{1k}x_{ik} + f_{ik}$
1	35	0	50	3	$y_{i1} = 35.0 + 5.0x_{i1} + f_{i1}$
	40	3			
	60	3			
	65	6			
2	36	1	54	4	$y_{i2} = 30.0 + 6.0x_{i2} + f_{i2}$
	44	4			
	64	4			
	72	7			
3	39	2	60	5	$y_{i3} = 25.0 + 7.0x_{i3} + f_{i3}$
	50	5			
	70	5			
	81	8			
4	44	3	68	6	$y_{i4} = 20.0 + 8.0x_{i4} + f_{i4}$
	58	6			
	78	6			
	92	9			

For η^2,

$$\eta^2 = \text{BSS/TSS} = 20/92 = 0.22 \tag{8.23}$$

The multilevel structure based on the assumption of an individual and group effect (only) specifies that the group-level slope d_{1g} should equal the sum of the two effects b_1 and b_2. That sum is $6.5 - 5.0 = 1.5$. The value of the group-level slope is 6.0. The population slope d_{1p} should equal the sum of b_1 plus a proportion (0.22) of b_2. That value is $6.5 - 5.0(0.22) = 5.4$. The population slope equals 6.4. Therefore the aggregate slopes are not explained by the individual-model effects, as they were above. The discrepancies illustrate the effect of ignoring the interaction effect when it exists.

The correct model specifies that the within-group slopes d_{1k} are a function of the group-mean variable $\bar{X}$. The separate equation is

$$\begin{aligned} d_{1k} &= b_1 + b_3\bar{x}_k + v_k \\ &= 2.0 + 1.0\bar{x}_k + v_k \end{aligned} \tag{8.24}$$

Substantively, the individual effect suggests that personal sales increase with personal competitiveness, perhaps through increased effort. The negative group effect suggests that rising levels of competitiveness adversely affect sales, perhaps through a decrease in cooperation. The interaction effect suggests that noncompetitive individuals are more adversely affected by rising levels of competitiveness than competitive individuals.

The single equation for this model is

$$\begin{aligned} y_{ik} &= b_0 + b_1 x_{ik} + b_2 \bar{x}_k + b_3 x_{ik} \bar{x}_k + e_{ik} \\ &= 50.0 + 2.0 x_{ik} - 5.0 \bar{x}_k + 1.0 x_{ik} \bar{x}_k + e_{ik} \end{aligned} \tag{8.25}$$

Aggregating Eq. (8.25) to the level of sales districts gives

$$\bar{y}_k = b_0 + b_1 \bar{x}_k + b_2 \bar{x}_k + b_3 \bar{x}_k^2 + \bar{e}_k \tag{8.26}$$

Collecting terms produces

$$\begin{aligned} \bar{y}_k &= b_0 + (b_1 + b_2) \bar{x}_k + b_3 \bar{x}_k^2 + \bar{e}_k \\ &= 50.0 + (2.0 - 5.0) \bar{x}_k + 1.0 \bar{x}_k^2 + \bar{e}_k \end{aligned} \tag{8.27}$$

Eq. (8.27) indicates that the group-level relationship ought to be a second-degree polynomial curve. The form of this curve is

$$\begin{aligned} \bar{y}_k &= d_{0g} + d_{1g} \bar{x}_k + d_{2g} \bar{x}_k^2 + e_k \\ &= 50.0 - 3.0 \bar{x}_k + 1.0 \bar{x}_k^2 + e_k \end{aligned} \tag{8.28}$$

The coefficients in Eq. (8.28) are obtained directly from the multiple regression of $\bar{y}_k$ on $\bar{x}_k$ and $\bar{x}_k^2$. They are equal to those obtained indirectly in Eq. (8.27). Therefore the aggregate-level curve has been explained by the underlying model of individual effects. Note that the simple group-level slope does not equal the sum of b_1 and b_2. However, that sum does equal the partial regression coefficient d_{1g}.

To reveal the relationship of the aggregate slopes d_{1g} and d_{1p} to each other and to the effects specified in this contextual model, the multilevel structure is derived as before. See Section 8.3 and Appendix H. This produces the equation

$$\begin{aligned} d_{1p} &= b_1(1 - \eta^2) + b_3 \bar{x}(1 - \eta^2) + d_{1g} \eta^2 \\ &= 2(0.78) + 1(4.5)(0.78) + 6.0(0.22) = 6.4 \end{aligned} \tag{8.29}$$

Eq. (8.29) indicates several things about the relationship between the population slope and the group-level slope, and between the population slope and the individual effect. The difference between the population and group-level slopes depends on the signs and magnitudes of the individual and interaction effects, the mean of X, and η^2. When all signs are positive, the population slope is smaller than the group-level slope. It is smaller in proportion to the η^2, the mean of X, and the coefficients b_1 and b_3. Similarly, when the signs are positive, the population slope is larger than the individual-level coefficient in proportion to the η^2, the mean of X, and the values of b_3 and d_{1g}.

The limiting cases for differences between the population and group-level slopes, and between the population slope and the individual-level coefficient, are fixed by η^2 and $(1 - \eta^2)$. For example, if η^2 is zero, the population slope is smaller than the group-level slope by the sum of the individual coefficient b_1 plus the interaction effect times the value of the mean of X $(b_3\bar{x})$.

Since the direct estimate of d_{1p} on the left-hand side of Eq. (8.29) equals the value based on the right-hand side, the model has been correctly specified and the population slope has therefore been explained in terms of individual behavior.

8.7 The Implications of Nonlinear Context Effects in the Explanation of Aggregate Relationships

Extensions of the basic model of contextual analysis to consider nonlinear context effects are discussed in Section 6.1. The general case specifies that within-group intercepts d_{0k} and slopes d_{1k} are nonlinear functions of the group-mean variable $\bar{X}$. In the sales-group study it might be postulated that the group effect associated with levels of competitiveness are stronger in the upper ranges of $\bar{X}$ than in the lower ranges. This would indicate that the adverse consequences of rising competitiveness

Figure 8.3
Hypothetical Curves Illustrating Group Intercepts and Slopes as Nonlinear Functions of Group Means for Sales-group Example

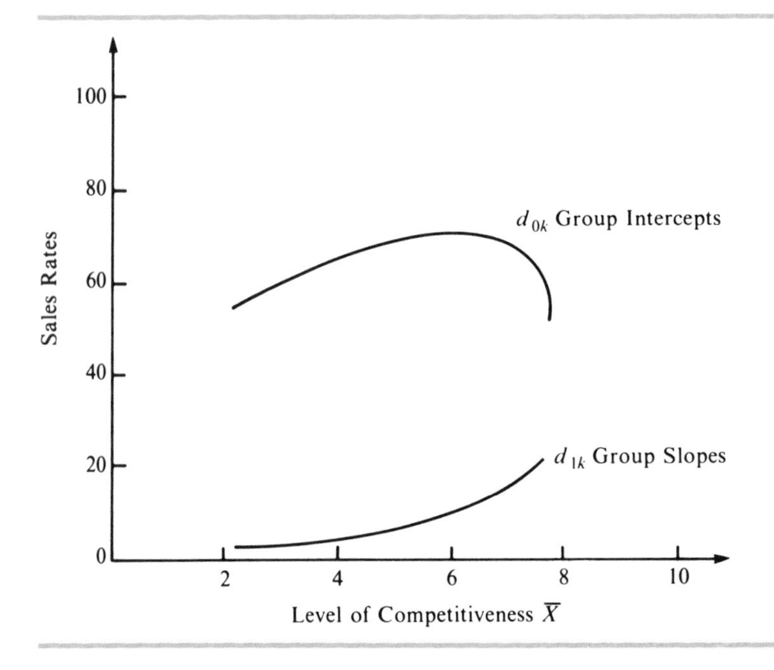

are not felt until a certain level of competitiveness is reached. A curve suggesting this phenomenon is shown in Figure 8.3.

Similarly, it could be that the personal effect of competitiveness on sales is strengthened by competitiveness levels at increasing rates. This would indicate that competitive individuals are less affected by competitiveness levels than noncompetitive individuals. A curve suggesting this phenomenon is also shown in Figure 8.3.

Curves relating the intercepts and slopes to the group means can take any form. However, a common form is a second-degree polynomial. To illustrate, another data set for the sales-group study is presented in Table 8.3. The separate equations for this model are

$$\begin{aligned} d_{0k} &= b_0 + b_2\bar{x}_k + b_4\bar{x}_k^2 + u_k \\ &= 50.0 + 3.0\bar{x}_k - 1.0\bar{x}_k^2 + u_k \end{aligned} \tag{8.30}$$

$$\begin{aligned} d_{1k} &= b_1 + b_3\bar{x}_k + b_5\bar{x}_k^2 + v_k \\ &= 6.0 + 0.0\bar{x}_k + 0.0\bar{x}_k^2 + v_k \end{aligned} \tag{8.31}$$

The coefficients in Eq. (8.31) indicate that there is neither an interaction effect nor a second-degree interaction effect. The single equation is

$$y_{ik} = b_0 + b_1x_{ik} + b_2\bar{x}_k + b_3x_{ik}\bar{x}_k + b_4\bar{x}_k^2 + b_5x_{ik}\bar{x}_k^2 + e_{ik} \tag{8.32}$$

Table 8.3
Data for the Hypothetical Study of Sales Groups Illustrating the Implications of Higher Degree Context Effects in the Explanation of Aggregate Relationships

DISTRICT NUMBER k	PERSONAL SALES y_{ik}	COMPETITIVENESS SCORES x_{ik}	DISTRICT AVERAGES $\bar{y}_k$	$\bar{x}_k$	WITHIN-DISTRICT REGRESSION LINES $y_{ik} = d_{0k} + d_{1k}x_{ik} + f_{ik}$
1	50	0	68	3	$y_{i1} = 50.0 + 6.0x_{i1} + f_{i1}$
	58	3			
	78	3			
	86	6			
2	52	1	70	4	$y_{i2} = 46.0 + 6.0x_{i2} + f_{i2}$
	60	4			
	80	4			
	88	7			
3	52	2	70	5	$y_{i3} = 40.0 + 6.0x_{i3} + f_{i3}$
	60	5			
	80	5			
	88	8			
4	50	3	68	6	$y_{i4} = 32.0 + 6.0x_{i4} + f_{i4}$
	58	6			
	78	6			
	86	9			

The group-level equation is obtained by aggregating Eq. (8.32) to the level of sales districts. This is

$$\bar{y}_k = b_0 + (b_1 + b_2)\bar{x}_k + (b_3 + b_4)\bar{x}_k^2 + b_5\bar{x}_k^3 + \bar{e}_k \tag{8.33}$$

Eq. (8.33) has the same form as a third-degree polynomial regression equation relating $\bar{Y}$ and $\bar{X}$. This would appear as an S curve in a graph. Such a relationship should not be observed in this example, because there is no second-degree interaction effect to produce that phenomenon at the aggregate level.

The group-level equation is

$$\bar{y}_k = d_{0g} + d_{1g}\bar{x}_k + d_{2g}\bar{x}_k^2 + d_{3g}\bar{x}_k^3 + e_k \tag{8.34}$$

The coefficients in Eq. (8.34) are denoted by d to distinguish them from the coefficients in Eq. (8.33). If the model of individual behavior is correctly specified, the coefficients in Eq. (8.33) should equal the coefficients in Eq. (8.34). These expected relationships are

$$d_{0g} = b_0, \qquad d_{1g} = (b_1 + b_2), \qquad d_{2g} = (b_3 + b_4), \qquad d_{3g} = b_5 \tag{8.35}$$

The values for the sales-group data based on Eq. (8.32) are

$$b_0 = 50.0, \qquad b_1 + b_2 = 9.0, \qquad b_3 + b_4 = -1.0, \qquad b_5 = 0.0 \tag{8.36}$$

The values obtained directly from Eq. (8.33) are 50.0, 9.0, -1.0, and 0.0. They are the same. Therefore, the aggregate nonlinear relationship observed in Eq. (8.33) has been explained in terms of individual behavior.

It is observed in Chapter 6 that the separate equation procedure for estimating a model of individual behavior will not produce the same results as the single equation procedure when existing higher degree effects are ignored. The implication for explaining aggregate relationships is that coefficients based on the incorrectly specified model of individual behavior will not reproduce the aggregate-level relationships. Comparing the single- and separate-equation results therefore serves as a check on the specification of the model of individual behavior.

The population slope estimate d_{1p} is 4.7. The simple group-level slope d_{1g} equals zero. The values of the individual effect b_1, the interaction effect b_3, and the second-degree interaction effect b_5 equal 6.0,0.0, and 0.0, respectively. η^2 is 0.22. As in previous discussions, the relationships between these components of the multilevel structure can be derived. This essentially involves a further partitioning of the population slope d_{1p} to incorporate the measures of nonlinear effects. Details follow from an extension of what is given in Appendix H. The result is

$$\begin{aligned} d_{1p} &= b_1(1 - \eta^2) + b_3\bar{x}(1 - \eta^2) + b_5\overline{x^2}(1 - \eta^2) + d_{1g}\eta^2 \\ &= 4.68 + 0.0 + 0.0 + 0.0 = 4.68 \end{aligned} \tag{8.37}$$

where $\overline{\bar{x}^2}$ is the weighted mean of $\bar{x}_k^2$ using η_k as weight, and d_{1g} is the coefficient for $\bar{x}_k$ in the linear relationship between the group means. When the model of individual behavior specified above is correct, the value of d_{1p}, obtained directly from the regression of Y on X, will equal the value obtained from the right-hand side of Eq. (8.37). This is true in the example. Therefore it can be said that the population slope has been explained in terms of individual behavior. The difference between the population slope, the group-level slope, and the individual effect depends on the specific shape of the curves observed in the separate equations [Eqs (8.30) and (8.31)].

In the literature it has been suggested that a curve in the relationship between group-mean variables indicates the presence of an interaction effect (cf. Hardor and Pappi, 1969). The example illustrates how a curve can be observed without the existence of an interaction effect. Instead, the consequence of a second-degree group effect is observed at the aggregate level.

8.8 *Summary*

The multilevel objective of understanding aggregate relationships in terms of individual behavior is examined in this chapter. Two aggregate relationships are studied. The first is the relationship between group-mean variables. The second involves the relationship between individual-level variables analyzed at the level of a population. Understanding these relationships requires the aggregation of correctly specified models of individual behavior. Generally, models of individual behavior which consider only individual-level variables are inadequate. The key is to consider explicitly the effects of context, mediated through the group-mean variable. The possibility that these effects involve a cross-level interaction or operate in nonlinear ways must also be considered. Beyond the explicit aim to explain aggregate relationships, the specification of multilevel relationships serves as a diagnostic aid for detecting incorrectly specified models of individual behavior. The concepts and procedures in this chapter provide a methodologically explicit step toward integrating micro and macro sociology. Additional substantive examples are presented in the next chapter.

Additional Examples of Explaining Aggregate Relationships

To further illustrate concepts and procedures for explaining aggregate relationships, three substantive examples are presented. The first concentrates on a population slope observed in a study of 172 book clubs and its relationship to the group-level slope, the measure of individual effect, and group variances (cf. Davis, 1961a). The second and third examples involve the analysis of illiteracy introduced in Section 6.1. The data from the 1930 census are the same used by Robinson (1950) in his analysis of ecological correlations and individual behavior. One example concerns the relationship between illiteracy and race. The other involves the relationship between illiteracy and nativity (foreign-born). These examples focus on the explanation of group-level relationships. One demonstrates how a straight-line relationship between group-mean variables is explained by nonlinear context effects. The other shows how a complex curve relating group-mean variables is explained by nonlinear context effects.

9.1 *Example 1: Multilevel Structure in the Analysis of Participation in Great Books Clubs*

The individual-level dependent variable Y in this analysis is whether a member stays or drops out of the book club. The variable X is whether a member is active or not active in meetings. The question concerns whether active members are less likely to drop out than inactive members. Data for seven groups of clubs (representing 172 clubs) are shown in Table 9.1. Dummy variables are defined for Y and X. Stayers and actives are given 1's, and dropouts and inactives are given 0's. The group-mean

Table 9.1
Great-book-club Data on Activity and Participation

Club Group 1

	Inactive	Active	
Stay	214	31	245
	(0.47)*	(0.66)	0.49
Drop	241	16	257
	(0.53)	(0.34)	0.51
	455	47	502
	0.91	0.09	1.00

$d_{01} = 0.47, \quad d_{11} = 0.19$

Club Group 2

	Inactive	Active	
Stay	69	30	99
	(0.55)	(0.68)	0.59
Drop	56	14	70
	(0.45)	(0.32)	0.41
	125	44	169
	(0.74)	0.26	1.00

$d_{02} = 0.55, \quad d_{12} = 0.13$

Club Group 3

	Inactive	Active	
Stay	108	74	182
	(0.60)	(0.70)	0.64
Drop	72	31	103
	(0.40)	(0.30)	0.36
	180	105	285
	0.63	0.37	1.00

$d_{03} = 0.60, \quad d_{13} = 0.10$

Club Group 4

	Inactive	Active	
Stay	80	77	157
	(0.59)	(0.71)	0.64
Drop	56	31	87
	(0.41)	(0.29)	0.36
	136	108	244
	0.56	0.44	1.00

$d_{04} = 0.59, \quad d_{14} = 0.12$

Club Group 5

	Inactive	Active	
Stay	78	108	186
	(0.69)	(0.81)	0.76
Drop	35	25	60
	(0.31)	(0.19)	0.24
	113	133	246
	(0.46)	(0.54)	1.00

$d_{05} = 0.69, \quad d_{15} = 0.12$

Club Group 6

	Inactive	Active	
Stay	36	82	118
	(0.63)	(0.82)	0.75
Drop	21	18	39
	(0.37)	(0.18)	0.25
	57	100	157
	0.36	0.64	1.00

$d_{06} = 0.63, \quad d_{16} = 0.19$

Club Group 7

	Inactive	Active	
Stay	17	80	97
	(0.68)	(0.78)	0.76
Drop	8	22	30
	(0.32)	(0.22)	0.24
	25	102	127
	0.20	0.80	1.00

$d_{07} = 0.68, \quad d_{17} = 0.10$

Sum Table

	Inactive	Active	
Stay	602	482	1084
	(0.55)	(0.75)	0.63
Drop	489	157	646
	(0.45)	(0.25)	0.37
	1091	639	1730
	0.63	0.37	1.00

$d_{0p} = 0.55, \quad d_{1p} = 0.20$

* Column conditional proportions in parentheses.
Source: Adapted from Davis, 1961b, p. 580.

variable $\bar{X}$ equals the proportion of actives and $\bar{Y}$ the proportion of stayers. For club group 1 in Table 9.1, these values are 0.09 and 0.49, respectively.

The within-group variances of X are denoted by s_k^2. They are obtained from $\bar{x}_k(1 - \bar{x}_k)$, involving the column margins in Table 9.1. For example, the variance of X in club 1 is $(0.09)(0.91) = 0.08$. The conditional proportion of stayers given active participation in meetings is designated m_{ak}, where k identifies the group number. For example, $m_{a1} = 0.66$ is the conditional proportion of stayers given active participation in club 1. The conditional proportion of stayers given inactivity in meetings is m_{nak}. In club 1 that value is 0.47.

The relationships between Y and X within book clubs is expressed in

$$y_{ik} = d_{0k} + d_{1k}x_{ik} + f_{ik} \tag{9.1}$$

Equivalences between conditional proportions in contingency tables and regression coefficients are described in Chapter 5. Accordingly, the intercepts d_{0k} equal the conditional proportions m_{nak}. For club 1 the value is 0.47. The values of the slopes d_{1k} equal the differences between the conditional proportions $m_{ak} - m_{nak}$. For club 1 this becomes 0.19. Table 9.1 indicates that the slopes relating staying and active participation range from 0.10 to 0.19.

The population or sum table for all the book clubs is the last table given in Table 9.1. Conditional proportions for the population are denoted by m_a and m_{na}. The equation for the population relationship is

$$y_{ik} = d_{0p} + d_{1p}x_{ik} + e_{ik} \tag{9.2}$$

The intercept in Eq. (9.2) is obtained from the conditional proportion m_{na}. The slope equals the difference in the conditional proportions. This gives

$$\begin{aligned} d_{0p} &= m_{na} = 0.55 \\ d_{1p} &= m_a - m_{na} = 0.75 - 0.50 = 0.20 \end{aligned} \tag{9.3}$$

The group-level relationship between proportions of stayers and proportions of active participators is specified in the equation

$$\begin{aligned} \bar{y}_k &= d_{0g} + d_{1g}\bar{x}_k + e_k \\ &= 0.48 + 0.41\bar{x}_k + e_k \end{aligned} \tag{9.4}$$

The focus of this example is on the population slope $d_{1p} = 0.20$ and its relationship to other parts of the multilevel structure. A contextual analysis to obtain measures of individual, group, and interaction effects is required. This analysis indicates the existence of individual and group effects. The interaction effect is negligible. The coefficients for this model of individual behavior are

$$\begin{aligned} d_{0k} &= b_0 + b_2\bar{x}_k + u_k \\ &= 0.47 + 0.29\bar{x}_k + u_k \end{aligned} \tag{9.5}$$

$$\begin{aligned} d_{1k} &= b_1 + v_k \\ &= 0.14 + v_k \end{aligned} \tag{9.6}$$

As before, b_1 is a measure of the individual effect, and b_2 is a measure of the group effect.

The multilevel structure for this problem is specified in Eqs. (8.14) through (8.17). All necessary coefficients for these equations have been obtained above except the value of η^2 which can be obtained from the table margins in Table 9.1. The within sum of squares of X is

$$\text{WSS} = \sum_k s_k^2 n_k = \sum_k \bar{x}_k(1 - \bar{x}_k)n_k = 319.1 \tag{9.7}$$

The total sum of squares is

$$\text{TSS} = s^2 n = \bar{x}(1 - \bar{x})n = 403.0 \tag{9.8}$$

The difference is the between sum of squares:

$$\text{BSS} = \text{TSS} - \text{WSS} = 403.0 - 319.1 = 83.9 \tag{9.9}$$

The ratio of the between sums of squares to the total sums of squares is

$$\eta^2 = \text{BSS}/\text{TSS} = 0.21 \tag{9.10}$$

The multilevel relationships can be observed by substituting the above estimates into Eqs. (8.14) through (8.17). This gives

$$\begin{aligned} d_{1g} &= b_1 + b_2 \\ 0.41 &\approx 0.14 + 0.29 = 0.43 \end{aligned} \tag{9.11}$$

$$\begin{aligned} d_{1p} &= b_1 + b_2\eta^2 \\ 0.20 &= 0.14 + 0.29(0.21) = 0.20 \end{aligned} \tag{9.12}$$

$$\begin{aligned} d_{1p} &= b_1(1 - \eta^2) + d_{1g}\eta^2 \\ 0.20 &= 0.14(0.79) + 0.41(0.21) = 0.20 \end{aligned} \tag{9.13}$$

$$\begin{aligned} d_{1p} &= d_{1g} - b_2(1 - \eta^2) \\ 0.20 &\approx 0.41 - 0.29(0.79) = 0.18 \end{aligned} \tag{9.14}$$

The values on the left of Eqs. (9.11) through (9.14) were obtained directly from the group-level regression and the population regression. The values on the right-hand side were obtained from the contextual model of individual behavior. Since they are close in value, the aggregate slopes d_{1g} and d_{1p} have been explained in terms of the specified model of individual behavior. Specifically, Eq. (9.11) indicates that the group-level slope is the consequence of individual and group effects. Eq. (9.12) indicates that the population slope is the sum of the individual effect plus a portion of the group effect. Since that portion (η^2) is relatively small, very little of the group effect goes toward the observed value of the population slope. Eq. (9.14) shows that the population slope is smaller than the group-level slope by 79 percent. These observations explain why the population slope in this problem is closer in value to the coefficient for the individual effect than to the group-level slope.

Effects of changing the between-group sum of squares in the selection and definition of groups are illustrated by this example. Seven groups were made from the 172 book clubs by grouping them on the proportions of active participators $\bar{x}_k$. This expectedly reduces the variation between groups. Since the total variation (obtained from the sum table) is unaffected, η^2 for the seven groups is smaller than for the 172 clubs. The consequences of reducing η^2 can be observed in Eqs. (9.11) through (9.14). Specifically, the population slope would be nearer the group-level slope and further from the coefficient for the individual effect. Since the value of the individual coefficient b_1 would be unaffected and the group-level slope would be increased, the group effect would be larger. This can be seen in Eq. (9.12). The likely consequence of grouping in this case is to underestimate the group effect.

9.2

Example 2: Explaining the Group-level Relationship between Illiteracy Rates and the Racial Composition of States

The emphasis in this example is to explain the group-level relationship observed between proportions of Blacks and proportions of illiterates across states. The variable X is defined to equal 1 for Blacks and 0 for non-Blacks. The variable Y is defined to equal 1 for illiterates and 0 for literates. This makes the group means $\bar{x}_k$ equal the state proportions of Blacks and the group means $\bar{y}_k$ equal the state proportions of illiterates. The rate of illiteracy for Blacks by state is denoted by m_{bk}. The rate for non-Blacks is m_{nbk}. These are the conditional proportions of illiterates for Blacks and for non-Blacks.

The within-state relationships between race and illiteracy are analyzed by the equation

$$y_{ik} = d_{0k} + d_{1k}x_{ik} + f_{ik} \tag{9.15}$$

The intercepts and slopes in Eq. (9.15) can be obtained directly from the illiteracy rates for Blacks and non-Blacks, that is,

$$d_{0k} = m_{nbk}, \qquad d_{1k} = m_{bk} - m_{nbk} \tag{9.16}$$

The numerical values of the slopes range from -0.07 to 0.22.

The group-level equation relating state proportions of Blacks and illiterates is

$$\begin{aligned} \bar{y}_k &= d_{0g} + d_{1g}\bar{x}_k + e_k \\ &= 0.02 + 0.25\bar{x}_k + e_k \end{aligned} \tag{9.17}$$

The emphasis of this analysis is to explain the slope in Eq. (9.17) in terms of individual behavior.

One way to proceed is to assume that the within-group slopes and intercepts in Eq. (9.15) vary only randomly across states. This gives

$$\begin{aligned} d_{0k} &= b_0 + u_k \\ &= 0.03 + u_k \end{aligned} \tag{9.18}$$

$$\begin{aligned} d_{1k} &= b_1 + v_k \\ &= 0.05 + v_k \end{aligned} \tag{9.19}$$

The model specified in Eqs. (9.18) and (9.19) is the special case of individual effect only described in Chapter 3. The same model can be expressed in terms of the state illiteracy rates for Blacks and non-Blacks by substituting d_{0k} and d_{1k} from Eqs. (9.18) and (9.19) into Eq. (9.16). This gives

$$\begin{aligned} m_{nbk} &= b_0 + u_k \\ &= 0.03 + u_k \end{aligned} \tag{9.20}$$

$$m_{bk} - m_{nbk} = b_1 + v_k = 0.05 + b_k \tag{9.21}$$

Substituting m_{nbk} from Eq. (9.20) into Eq. (9.21) gives

$$\begin{aligned} m_{bk} &= (b_1 + b_0) + (v_k + u_k) \\ &= (0.05 + 0.03) + (v_k + u_k) \\ &= 0.08 + (v_k + u_k) \end{aligned} \tag{9.22}$$

Eqs. (9.20) and (9.22) show that the assumption of an individual effect only is equivalent to assuming that the illiteracy rates for Blacks and non-Blacks are unaffected by the racial composition of states. A graph of this model appears in Figure 9.1. The regression lines relating the illiteracy rates to the proportions of Blacks are parallel and have zero slopes.

The single equation for this model is

$$y_{ik} = b_0 + b_1 x_{ik} + e_{ik} \tag{9.23}$$

Aggregating the observations for the variables in Eq. (9.23) to the level of states produces

$$\bar{y}_k = b_0 + b_1\bar{x}_k + \bar{e}_k \tag{9.24}$$

Figure 9.1
Relationships between Illiteracy Rates and State Proportions of Blacks under the Assumption of Individual Effects Only

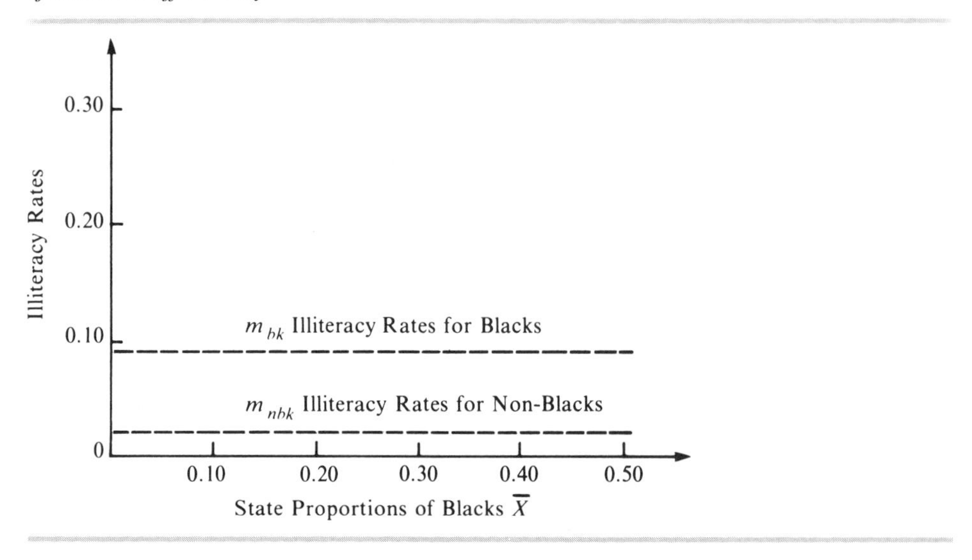

Substituting b_0 and b_1 from Eqs. (9.18) and (9.19) into Eq. (9.24) gives

$$\bar{y}_k = 0.03 + 0.05\bar{x}_k + \bar{e}_k \tag{9.25}$$

If the model specifying the individual effect only is correct, the coefficients in Eq. (9.25) should equal the coefficients obtained directly from the group-level regression in Eq. (9.17). Compare 0.02 to 0.03 and 0.25 to 0.05. They are not the same. Also note that Eq. (9.23) is the same form as the population equation. Coefficients obtained directly from this regression are 0.55 and 0.20. They also do not correspond to the estimates of b_0 and b_1. Therefore the model of individual behavior specifying only an individual effect is incorrect, and the group-level relationship is not adequately explained in terms of individual behavior.

Instead of the individual effect only model, consider the basic model of contextual analysis. Equations for this model are

$$\begin{aligned} d_{0k} &= b_0 + b_2\bar{x}_k + u_k \\ &= 0.03 + 0.04\bar{x}_k + u_k \end{aligned} \tag{9.26}$$

$$\begin{aligned} d_{1k} &= b_1 + b_3\bar{x}_k + v_k \\ &= 0.01 + 0.44\bar{x}_k + v_k \end{aligned} \tag{9.27}$$

The single equation is

$$y_{ik} = b_0 + b_1 x_{ik} + b_2\bar{x}_k + b_3 x_{ik}\bar{x}_k + e_{ik} \tag{9.28}$$

This model can be expressed in terms of the relationships between the illiteracy rates for Blacks and non-Blacks and the proportions of Blacks by substituting d_{0k} and d_{1k} from Eqs. (9.26) and (9.27) into Eq. (9.16). Thus,

$$m_{nbk} = b_0 + b_2\bar{x}_k + u_k \tag{9.29}$$

$$m_{bk} - m_{nbk} = b_1 + b_3\bar{x}_k + v_k \tag{9.30}$$

Moving m_{nbk} to the right-hand side of Eq. (9.30) and substituting m_{nbk} from Eq. (9.29) gives

$$m_{bk} = (b_0 + b_1) + (b_2 + b_3)\bar{x}_k + (v_k + u_k) \tag{9.31}$$

Eqs. (9.29) and (9.31) specify that the illiteracy rates for Blacks and non-Blacks are functions of the state proportions of Blacks. The coefficients are shown in the equation

$$m_{bk} = 0.04 + 0.48\bar{x}_k + (v_k + u_k) \tag{9.32}$$

$$m_{nbk} = 0.03 + 0.04\bar{x}_k + u_k \tag{9.33}$$

These relationships are shown in Figure 9.2. The small value of the lower regression slope (0.04) indicates that illiteracy rates among non-Blacks are only slightly affected by racial composition. Therefore there is no substantial group effect. The two regression lines intersect the vertical axis at nearly the same point. Therefore there is no substantial individual effect of being Black. However, the slope of the

Figure 9.2
Expected Relationships between Illiteracy Rates and State Proportions of Blacks under the Assumption of Individual, Group, and Interaction Effects

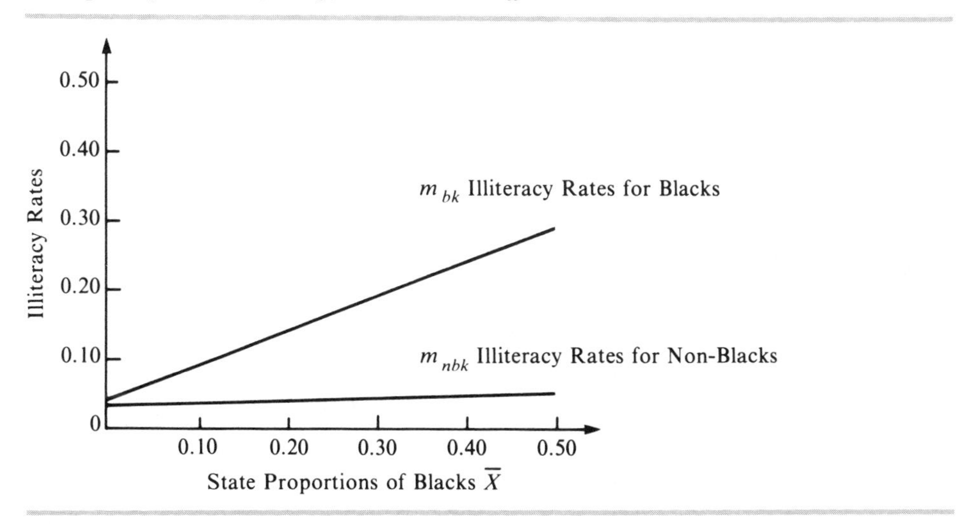

upper regression line (0.48) shows that illiteracy rates among Blacks increase rather markedly with increasing proportions of Blacks. This indicates an interaction effect between color as a property of individuals and color as a property of states. Substantively, this likely reflects racial inequalities in the quality and availability of public education at the time of the 1930 census.

Aggregating the single equation [Eq. (9.23)] to the level of states gives

$$\bar{y}_k = b_0 + (b_1 + b_2)\bar{x}_k + b_3\bar{x}_k^2 + \bar{e}_k \tag{9.34}$$

Substituting for the coefficients from Eqs. (9.26) and (9.27) into Eq. (9.34) gives

$$\bar{y}_k = 0.03 + 0.05\bar{x}_k + 0.44\bar{x}_k^2 + \bar{e}_k \tag{9.35}$$

The coefficients associated with the squared term in Eq. (9.35) suggest that a substantial second-degree polynomial curve should be observed in the relationship between the proportions of Blacks and proportions of illiterates. Instead, a direct estimate of the group-level equation containing a squared term gives

$$\bar{y}_k = 0.02 + 0.25\bar{x}_k + 0.01\bar{x}_k^2 + e_k \tag{9.36}$$

The result barely shows a trace of the expected curve. Therefore the model of individual behavior is not yet correctly specified.

A closer examination of the relationships between the illiteracy rates for Blacks and non-Blacks and state proportions of Blacks shown as straight lines in Figure 9.2 reveals that they are actually nonlinear. The illiteracy rates for Blacks m_{bk} flatten where the proportions of Blacks are high. The illiteracy rates for non-Blacks m_{nbk} are highest in the middle ranges of racial composition. These relationships are shown in Figure 9.3. The equations for second-degree polynomials are

$$\begin{aligned} m_{bk} &= c_1 + c_2\bar{x}_k + c_3\bar{x}_k^2 + e_{bk} \\ &= 0.03 + 0.68\bar{x}_k - 0.42\bar{x}_k^2 + e_{bk} \end{aligned} \tag{9.37}$$

$$\begin{aligned} m_{nbk} &= c_4 + c_5\bar{x}_k + c_6\bar{x}_k^2 + e_{nbk} \\ &= 0.02 + 0.23\bar{x}_k - 0.44\bar{x}_k^2 + e_{nbk} \end{aligned} \tag{9.38}$$

Expressed in terms of state intercepts and slopes, the coefficients are shown in the equations

$$\begin{aligned} d_{0k} &= b_0 + b_2\bar{x}_k + b_4\bar{x}_k^2 + u_k \\ &= 0.02 + 0.23\bar{x}_k - 0.44\bar{x}_k^2 + u_k \end{aligned} \tag{9.39}$$

$$\begin{aligned} d_{1k} &= b_1 + b_3\bar{x}_k + b_5\bar{x}_k^2 + v_k \\ &= 0.01 + 0.45\bar{x}_k + 0.02\bar{x}_k^2 + v_k \end{aligned} \tag{9.40}$$

Figure 9.3
Relationships between Illiteracy Rates and State Proportions of Blacks with Nonlinear Context Effects

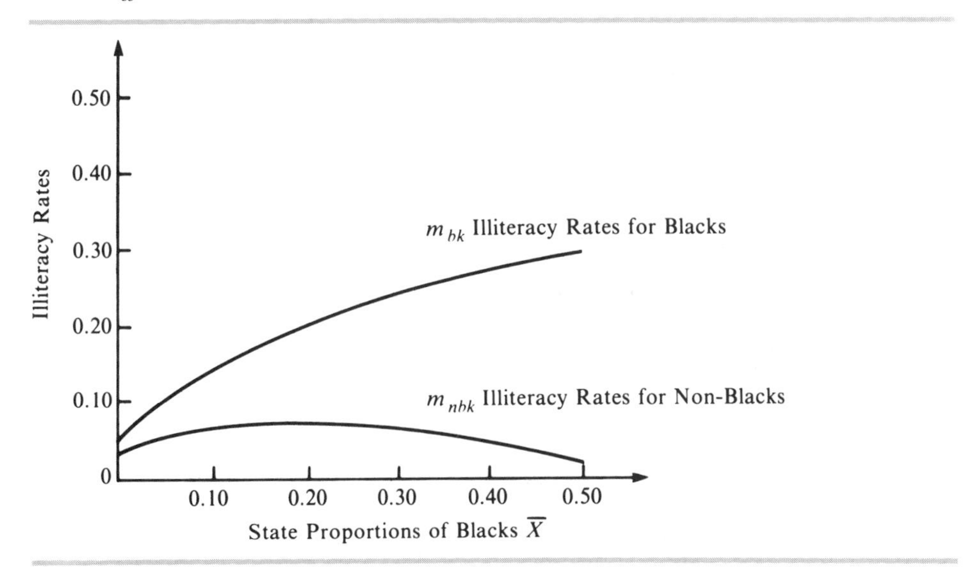

These results indicate the existence of a second-degree group effect, but essentially no second-degree interaction effect. The single equation for this model is

$$y_{ik} = b_0 + b_1 x_{ik} + b_2 \bar{x}_k + b_3 x_{ik} \bar{x}_k + b_4 \bar{x}_k^2 + e_{ik} \tag{9.41}$$

Aggregating this equation to the level of states gives

$$\bar{y}_k = b_0 + (b_1 + b_2)\bar{x}_k + (b_3 + b_4)\bar{x}_k^2 + \bar{e}_k \tag{9.42}$$

Substituting the estimates from Eqs. (9.39) and (9.40) into Eq.(9.42) gives

$$\begin{aligned} \bar{y}_k &= 0.02 + (0.01 + 0.23)\bar{x}_k + (0.45 - 0.44)\bar{x}_k^2 + \bar{e}_k \\ &= 0.02 + 0.24\bar{x}_k + 0.01\bar{x}_k^2 + \bar{e}_k \end{aligned} \tag{9.43}$$

The results from the model of individual behavior are now the same as from the regression in Eq. (9.36). Therefore the group-level relationship between proportions of illiterates and proportions of Blacks, which is virtually a straight line, has been explained in terms of a model of individual-behavior containing nonlinear context effects. Specifically, the first coefficient is the consequence of individual and group effects operating in the same direction. The second coefficient is the consequence of an interaction effect and a second-degree group effect operating in opposite directions.

The population slope is obtained from

$$
\begin{aligned}
y_{ik} &= d_{0p} + d_{1p}x_{ik} + e_{ik} \\
&= 0.03 + 0.13x_{ik} + e_{ik}
\end{aligned} \tag{9.44}
$$

The interpretation of the population slope in terms of multilevel structure can be observed with Eq. (8.29). η^2 equals 0.40, and the grand mean $\bar{x}$ is 0.09. The equation is

$$
\begin{aligned}
d_{1p} &= b_1(1 - \eta^2) + b_3\bar{x}(1 - \eta^2) + d_{1g}\eta^2 \\
0.13 &= 0.01(0.60) + 0.45(0.09)(0.60) + 0.25(0.40) = 0.13
\end{aligned} \tag{9.45}
$$

Eq. (9.45) shows that the population relationship between illiteracy and race is made up primarily of a part associated with the group-level slope and a part associated with the interaction effect.

9.3

Example 3: Explaining the Group-level Relationship between Illiteracy and Nativity

In the previous example effects cancel out to explain a straight-line relationship between group variables. The relationship between illiteracy and nativity provides an example where effects do not cancel. The consequence observed at the group level is a complex curve. This example illustrates the explanation of that curve in terms of individual behavior.

The dependent variable Y is defined to equal 1 for illiterates and 0 for literates. The variable X is defined to equal 1 for foreign birth and 0 for native birth. The group-mean variable $\bar{X}$ represents the state proportions of foreign-born. The mean variable $\bar{Y}$ represents the state proportions of illiterates. State illiteracy rates are designated m_{fk} for the foreign-born and m_{nk} for the native-born. These are the conditional proportions of illiterates given foreign birth and native birth.

Within-state relationships between illiteracy and nativity are analyzed by the regression model

$$
y_{ik} = d_{0k} + d_{1k}x_{ik} + f_{ik} \tag{9.46}
$$

The intercepts d_{0k} in Eq. (9.46) equal the illiteracy rates for the native-born m_{nk}. The slopes d_{1k} equal the difference between the illiteracy rates $m_{fk} - m_{nk}$. The population (nation) relationship is observed in

$$
\begin{aligned}
y_{ik} &= d_{0p} + d_{1p}x_{ik} + e_{ik} \\
&= 0.04 + 0.06x_{ik} + e_{ik}
\end{aligned} \tag{9.47}
$$

This indicates a relatively weak but positive relationship between illiteracy and nativity at the national level.

Figure 9.4 shows a plot of the group-level relationship between proportions of foreign-born and proportions of illiterates. This plot indicates that the relationship between illiteracy and nativity across states is negative. Here is the anomaly of the group-level relationship with a sign opposite to the population relationship. Also striking is the shape of the group-level relationship. In states with relatively small proportions of foreign-born, the relationship is sharply negative. In states with relatively large numbers of foreign-born, the relationship is positive. In states with large proportions of foreign-born, the relationship begins to reverse again. The question is how this curve is explained in terms of multilevel effects.

Figure 9.4
Scatter Plot of the Relationship between Proportions of Illiterates and Proportions of Foreign-born

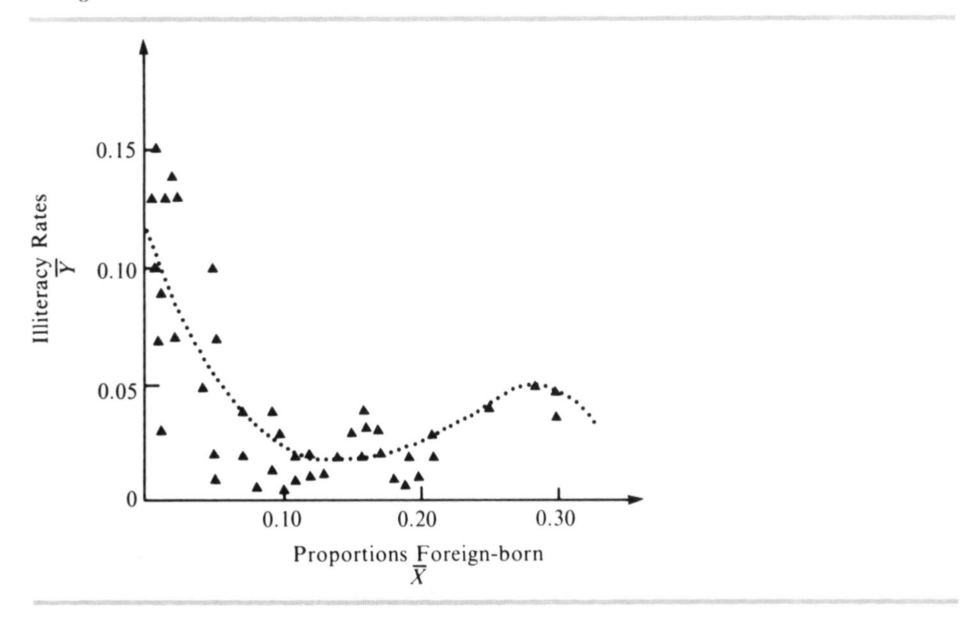

The curve in Figure 9.4 suggests an equation of the form

$$\begin{aligned} \bar{y}_k &= d_{0g} + d_{1g}\bar{x}_k + d_{2g}\bar{x}_k^2 + d_{3g}\bar{x}_k^3 + e_k \\ &= 0.11 - 1.79\bar{x}_k + 10.78\bar{x}_k^2 - 18.60\bar{x}_k^3 + e_k \end{aligned} \tag{9.48}$$

The fit of this curve is reflected in the changes of R^2 with the successive introduction of second- and third-degree terms. The increase was from 0.26 for the bivariate regression to 0.59 for the second-degree polynomial to 0.69 for the third-degree polynomial in Eq. (9.48).

The objective of a contextual analysis is to determine how the illiteracy rates for the foreign- and native-born are affected by the state proportions of foreign-born.

In terms of broad patterns, it is found that the relationship between the illiteracy rates for the native-born m_{nk} and proportions of foreign-born $\bar{x}_k$ is negative. This likely reflects that states with larger proportions of foreign-born tended to be states with better quality or more available education. For the foreign-born, the relationship between illiteracy rates and proportions of foreign-born is negligible. This means that the differences between illiteracy rates for the native- and foreign-born increase with the proportions of foreign-born. The interaction effect is likely a reflection of inequalities in availability or access to public education across states.

However, the basic model of contextual analysis implied above could not have produced the complex group-level curve observed in Figure 9.4. Nonlinear context effects are involved in the relationships between illiteracy rates and the proportions of foreign-born. These are observed in Figure 9.5. Illiteracy rates for the foreign- and native-born diminish in the lower ranges of the proportions of foreign-born. They increase in the middle ranges and decrease again in the upper ranges. An overall interaction effect is observed because the two curves change at different rates.

Figure 9.5
Relationships between Illiteracy Rates and State Proportions of Foreign-born

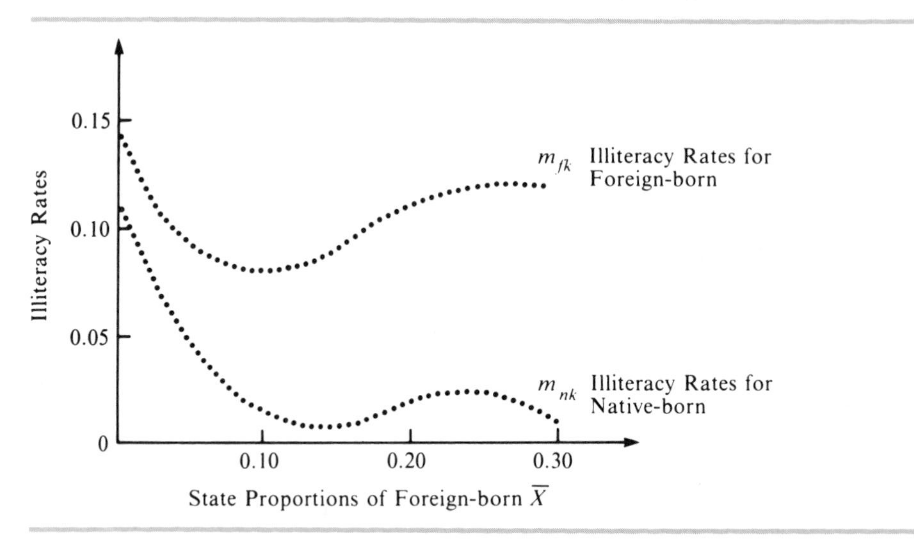

When this contextual model is expressed in terms of within-group intercepts and slopes, the following coefficients are obtained:

$$
\begin{aligned}
d_{0k} &= b_0 + b_2\bar{x}_k + b_4\bar{x}_k^2 + b_6\bar{x}_k^3 + u_k \\
&= 0.11 - 1.82\bar{x}_k + 10.40\bar{x}_k^2 - 18.15\bar{x}_k^3 + u_k
\end{aligned} \tag{9.49}
$$

$$
\begin{aligned}
d_{1k} &= b_1 + b_3\bar{x}_k + b_5\bar{x}_k^2 + v_k \\
&= 0.03 + 0.39\bar{x}_k - 0.45\bar{x}_k^2 + v_k
\end{aligned} \tag{9.50}
$$

The single equation for this model is

$$y_{ik} = b_0 + b_1 x_{ik} + b_2 \bar{x}_k + b_3 x_{ik}\bar{x}_k + b_4 \bar{x}_k^2 + b_5 x_{ik}\bar{x}_k^2 + b_6 \bar{x}_k^3 + e_{ik} \tag{9.51}$$

Aggregating Eq. (9.51) to the level of states produces

$$\bar{y}_k = b_0 + (b_1 + b_2)\bar{x}_k + (b_3 + b_4)\bar{x}_k^2 + (b_5 + b_6)\bar{x}_k^3 + \bar{e}_k \tag{9.52}$$

Substituting the estimates obtained from Eqs. (9.49) and (9.50) into Eq. (9.52) gives

$$\begin{aligned} \bar{y}_k &= 0.11 + (0.03 - 1.82)\bar{x}_k + (0.39 + 10.40)\bar{x}_k^2 - (0.45 + 18.15)\bar{x}_k^3 + \bar{e}_k \\ &= 0.11 - 1.79\bar{x}_k + 10.79\bar{x}_k^2 - 18.60\bar{x}_k^3 + \bar{e}_k \end{aligned} \tag{9.53}$$

The coefficients obtained from the contextual analysis and shown in Eq. (9.53) are the same as those obtained directly from the group-level regression [Eq. (9.48)]. Therefore it can be argued that the complex group-level relationship has been explained in terms of individual behavior. Specifically, the value of the first coefficient in Eq. (9.48) is primarily the consequence of the group effect. The second coefficient is in large part the consequence of the second-degree group effect. The last coefficient is primarily the result of the third-degree group effect.

As in previous examples, the population slope could be interpreted by deriving the multilevel structure for the model. Given the slopes obtained above, this additionally requires only η^2 and the grand mean $\bar{x}$.

9.4 *Summary*

This chapter further illustrates the concepts and procedures for explaining aggregate relationships introduced in Chapter 2 and elaborated in Chapter 8. Three substantive examples are examined. The first considers the explanation of the population slope observed in a study of 172 book clubs. The second and third examples illustrate the explanation of complex group-level relationships observed in an analysis of illiteracy, race, and nativity.

IV The Data Problem and Multilevel Analysis with Incomplete Data

Previous discussion assumed that complete data are available for each group involved in a multilevel analysis. With categorical data, this refers to marginal and joint distributions. With continuous data, this refers to group means, variances, and covariances. Frequently all of these data are not available or are impractical to acquire. Group data, sample population data, or data for one or two groups may be all that are available. Chapter 10 identifies the implications of incomplete data and the conditions under which different data sources may be combined to complete a multilevel analysis. Procedures for estimating individual, group, and interaction effects with various configurations of data are examined in Chapter 11. Chapter 12 presents additional substantive examples of these procedures. Suggestions for computer assistance are contained in Appendix A.

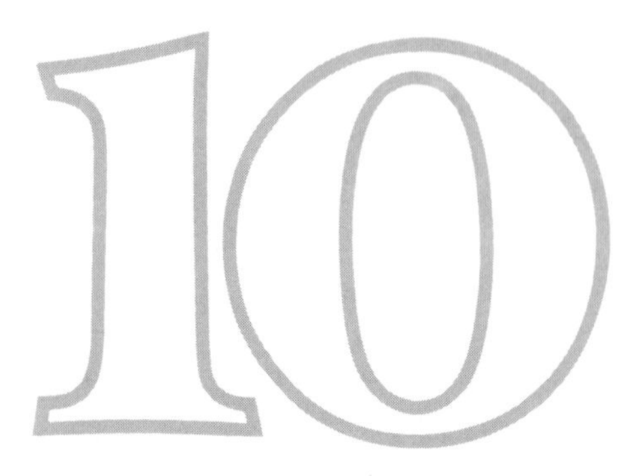

The Data Problem and Procedures with Incomplete Data

Two objectives of multilevel analysis are discussed in previous chapters. The first is to explain individual behavior in terms of individual and context variables. The second is to explain aggregate relationships in terms of contextual models of behavior. Corresponding to each objective is a set of equations containing parameters which are estimated from data. The first set is for the basic model of contextual analysis.

$$\begin{aligned} y_{ik} &= d_{0k} + d_{1k}x_{ik} + f_{ik} \\ d_{0k} &= b_0 + b_2\bar{x}_k + u_k \\ d_{1k} &= b_1 + b_3\bar{x}_k + v_k \\ y_{ik} &= b_0 + b_1x_{ik} + b_2\bar{x}_k + b_3x_{ik}\bar{x}_k + e_{ik} \end{aligned} \tag{10.1}$$

These equations contain coefficients for individual, group, and interaction effects on individual behavior.

The basic model of multilevel structure for explaining aggregate relationships introduced in Chapter 8 has the equations

$$\begin{aligned} d_{1p} &= b_1 + b_2\eta^2 \\ d_{1p} &= d_{1g} - b_2(1 - \eta^2) \\ d_{1p} &= b_1(1 - \eta^2) + b_3\bar{x}(1 - \eta^2) + d_{1g}\eta^2 \qquad \text{(with interaction)} \end{aligned} \tag{10.2}$$

These equations show how the measures of effects in the contextual analysis are related to the bivariate relationships observed at the group level and the population level. Derivations of these equations are discussed in Appendix H.

The purpose of this chapter is to make explicit what data are needed to estimate the parameters in the above equations. This requires a discussion of different types of single-source data. The problems of making individual-level inferences when complete data are not available are identified, and the fallacies of substituting one type of data for another are discussed. The possibility of combining data from different single sources to cope with the data problem is then introduced. A discussion of computer assistance for this problem in multilevel analysis is contained in Appendix A, Illustration 10.

10.1 *Types of Data and the Data Problem*

10.1.1. Types of Single-source Data

To visualize types of single-source data, refer to Table 10.1. This table contains data for the study of deviance (getting drunk) and social norms about deviance. These data are used in Chapter 5 to describe contextual analysis with contingency tables. There are five groups of colleges designated by the letter k (in subsequent discussions, the term "colleges" refers to groups of colleges). The dependent variable Y is defined to equal 1 for getting drunk and 0 for not getting drunk. The variable X equals 1 when the individual opposes deviance and 0 when the individual does not oppose deviance.

The group-mean variable $\bar{X}$ designates the proportions of individuals who oppose deviance in each college. The mean variable $\bar{Y}$ designates the proportions of individuals in each college who got drunk. The within-college variances s_{xk}^2 and s_{yk}^2 equal $\bar{x}_k(1-\bar{x}_k)$ and $\bar{y}_k(1-\bar{y}_k)$. These terms summarize the *marginal distribution* of X and the *marginal distribution* of Y.

The conditional proportions getting drunk among those opposing deviance in each college are designated m_{11k}. The conditional proportions getting drunk among those not opposed is m_{12k}. These observations appear in the *cells* of the tables in Table 10.1. Together, they summarize the *joint distribution* of X and Y.

The last table of Table 10.1 contains the observations on X and Y summed over all colleges. It is assumed that there are only the five colleges in the state. Therefore the data in that table refer to the state. The sum table contains the marginal distributions of X and Y for the population as a whole. These consist of the population proportions $\bar{x}$ and $\bar{y}$, and the population variances s_x^2 and s_y^2. The cells of the sum table contain the joint distribution of X and Y for the population. This involves the population conditional proportions m_{11} and m_{12}.

How much of the data described in Table 10.1 are necessary to complete a multilevel analysis of this problem? The answer is in the two sets of equations above. Eq. (10.1) indicates that one must be able to obtain within-college slopes and intercepts. From earlier discussion it is known that the intercepts d_{0k} equal the conditional proportions m_{11k}, and that the slopes d_{1k} equal the differences between the conditional proportions $m_{12k}-m_{11k}$. The within-group joint distributions must therefore be available. Eq. (10.1) indicates that the within-group marginal distributions of X

Table 10.1
Data for Five Colleges and Sum Table

GETTING DRUNK Y — DISAPPROVAL OF GETTING DRUNK X

College 1

Y	X = 0	X = 1	
1	561	15	576
	(0.731)	(0.122)	0.647
0	206	108	314
	(0.269)	(0.878)	0.353
	767	123	890
	0.862	0.138	

College 2

Y	X = 0	X = 1	
1	476	42	518
	(0.683)	(0.111)	0.481
0	221	338	559
	(0.317)	(0.889)	0.519
	697	380	1077
	0.647	0.353	

College 3

Y	X = 0	X = 1	
1	525	80	605
	(0.558)	(0.080)	0.312
0	416	920	1336
	(0.442)	(0.920)	0.688
	941	1000	1941
	0.485	0.515	

College 4

Y	X = 0	X = 1	
1	125	27	152
	(0.425)	(0.040)	0.158
0	169	644	813
	(0.575)	(0.960)	0.842
	294	671	965
	0.305	0.695	

College 5

Y	X = 0	X = 1	
1	18	4	22
	(0.360)	(0.010)	0.051
0	32	379	411
	(0.640)	(0.990)	
	50	383	433
	0.115	0.885	

Sum Table

Y	X = 0	X = 1	
1	1705	168	1873
	(0.620)	(0.066)	0.353
0	1044	2389	3433
	(0.380)	(0.934)	0.647
	2749	2557	5306
	0.518	0.482	1.000

must be available. This implies that the marginal distribution of the means $\bar{x}_k$ must be available. Eq. (10.2) contains the term d_{1g}. This comes from the regression of the within-group means of Y on the within-group means of X. Therefore the within-group marginal distributions of both X and Y are necessary. This implies that the joint distribution of $\bar{x}_k$ and $\bar{y}_k$ is required. Eq. (10.2) contains η^2, which requires the within-group marginal distributions of X and the population marginal distribution of X. That equation also contains the term d_{1p}, which is the population slope relating X and Y. The population joint distribution of X and Y is required to obtain d_{1p}.

The most severe interpretation of these equations is that all of the data in Table 10.1 are essential for multilevel analysis. The only time this data requirement would be met is with a complete enumeration of the population, as in a complete census. However, the specific information required are the within-group joint distributions of X and Y and the joint distribution of the means $\bar{x}_k$ and $\bar{y}_k$. These can be *estimated* from samples. Therefore the minimum data requirement for multilevel analysis is a representative sample of groups with representative samples of individuals within each group. This is referred to as *complete data.*

The complete data requirement in multilevel analysis often is impossible or costly to meet. The more common situation is to have *part* of the complete data. There are several possibilities. To illustrate, imagine that a survey of students had been completed in just one of the colleges. Respondents presumably would have been asked if they had gotten drunk and if they opposed deviance. This means that the marginal and joint distributions of X and Y are available for this one group. For subsequent reference, this kind of data is referred to as *group specific data.* See Table 10.2 for a description of group specific data. In this situation all the information for college 1 of Table 10.1 would appear, but all the data in the remaining tables would be missing. A multilevel analysis of the relationship between X and Y is impossible. With data for only one group there are only one group slope and one intercept and two means. Therefore it is impossible to measure individual and group effects in the explanation of getting drunk.

Table 10.2
Types of Single-source Data

Group specific data	Requires at least a random sample giving the joint distributions of individual-level variables in one of the groups making up the population.
Sample population data	Requires at least a random sample giving the joint distributions of individual variables making up the population.
Group data	Requires at least a random sample giving the distribution of the means or proportions of individual-level variables for groups making up the population.
Complete data	Requires at least a random sample of groups making up the population with group specific data for those groups.

A related possibility is that the study was conducted in two or more colleges. As this number increases and to the extent that groups are randomly selected, the complete data situation is approached.

Another possibility is that the whole state had been studied with a *sample* survey of the college population. Responses to presumably the same questions would provide estimates of the marginal and the joint distributions of X and Y at the level of the population. Such information appears in the sum table of Table 10.1. However, unless the sample had been explicity designed to provide group specific data, the within-group marginal and joint distributions in the other tables will be missing. As in the case of group specific data for one group, inferences cannot be made about individual and group effects in the explanation of getting drunk. For subsequent reference this data type is referred to as *sample population data*.

A third important possibility is that only the marginal information for each table in the deviance study was available. Perhaps a study of colleges had been carried out and only percentages or proportions were reported. These data provide the marginal distributions for X and Y and the distributions for $\bar{X}$ and $\bar{Y}$. Group data only might also occur when information on the same colleges comes from different studies. In this case, too, the marginal distributions are known and the joint distributions are not. Since the groups (colleges) are presumably the same for $\bar{y}_k$ and $\bar{x}_k$, their joint distribution is known. This data situation would be observed in Table 10.1 if the cells in all of the tables, including the population table, were empty. For subsequent reference, data of this type are referred to as *group data*. As in previous cases, Eqs. (10.1) and (10.2) show that a multilevel analysis of getting drunk and social norms about deviance is impossible with group data alone.

From the perspective of multilevel analysis, all three common data situations described above involve *incomplete data*. The important possibility of having various combinations of data types is the major theme of subsequent discussion.

10.1.2. The Problem of Cross-level Inference[1]

A contextual analysis with complete data for the college example produces these coefficients:

$$\begin{aligned} d_{0k} &= b_0 + b_2\bar{x}_k + u_k \\ &= 0.84 - 0.55\bar{x}_k + u_k \end{aligned} \tag{10.3}$$

$$\begin{aligned} d_{1k} &= b_1 + b_3\bar{x}_k + v_k \\ &= -0.68 + 0.40\bar{x}_k + v_k \end{aligned} \tag{10.4}$$

Coefficients for the group-level equation are

$$\begin{aligned} \bar{y}_k &= d_{0g} + d_{1g}\bar{x}_k + e_k \\ &= 0.76 - 0.85\bar{x}_k + e_k \end{aligned} \tag{10.5}$$

Given a nonzero value of the interaction effect b_3 in Eq. (10.4), the relationship between the group variables should be a second-degree polynomial. Coefficients for

[1] For a list of references directly or indirectly pertinent to the problem of cross-level inference, see footnote 1, Chapter 2.

this group-level equation are

$$\bar{y}_k = d_{0g} + d_{1g}\bar{x}_k + d_{2g}\bar{x}_k^2 + e_k$$
$$= 0.79 - 1.02\bar{x}_k + 0.16\bar{x}_k^2 + e_k \quad (10.6)$$

Estimates for the population equation are

$$y_{ik} = d_{0p} + d_{1p}x_{ik} + e_{ik}$$
$$= 0.62 - 0.55x_{ik} + e_{ik} \quad (10.7)$$

The population mean $\bar{x}$ equals 0.48 and η^2 equals 0.18

Substituting the above values into the last equation in Eq. (10.2) shows the relationships between parts of the multilevel structure. This gives

$$-0.55 = -0.68(1 - 0.18) + 0.40(0.48)(1 - 0.18) + (-0.85)(0.18)$$
$$-0.55 = -0.56 + 0.16 - 0.15 \quad (10.8)$$

Imagine that only group data are available to study the relationship between getting drunk and opposition to deviance. Only the margins in Table 10.1 are available to infer the missing joint distributions for the cells of the tables. The Duncan/Davis technique introduced in Chapter 2 was developed for this purpose (cf. Duncan and Davis, 1953). That method uses the marginal data to calculate the minimum and maximum values which the cell entries could take while still adding up to the margins. The problem with that method is that the range of possible within-group relationships increases as marginal proportions split more evenly, and are often so large as to be of little use.

Another attempt to recover group specific data from group data came to be known as *ecological* regression (cf. Goodman, 1953, 1959; Stokes, 1969). This approach assumes that within-group conditional proportions are constant across groups. This is equivalent to the assumption that within-group intercepts and slopes are constant. Under this assumption these constants will equal the intercept and slope obtained from the marginal data. Coefficients from the group-level regression are 0.76 and −0.83. The within-group conditional proportions are obtained from

$$m_{11k} = d_{0k} = 0.76 \quad (10.9)$$

$$m_{12k} = d_{0k} + d_{1k} = 0.76 - 0.83 = -0.07 \quad (10.10)$$

The value of m_{12k} inferred from this procedure is clearly incorrect since conditional proportions cannot be negative. Even the assumption that it equals zero is implausible in the substantive problem.

Eq. (10.2) shows that the procedure referred to as ecological regression is an attempt to substitute information from one single-source data type for information of a different type. In this instance the value of the population slope was inferred directly from the group-level slope. Eq. (10.2) shows that d_{1p} will equal d_{1g} only when b_2 and b_3 equal zero. The inference from group data to population data will be valid only when there are no group or interaction effects. An invalid inference of this type has become known as the *ecological fallacy*.

Substituting group-level information for population information represents only one kind of cross-level inference. Group data may be all that are available to estimate the slope in a particular group. For example, the inference may be from the group-level slope d_{1g} from Eq. (10.5) to the slope d_{15} for college 5 from Table 10.1. The two coefficients equal -0.83 and -0.35, respectively. Because they are different, one cannot be used in place of the other. Erroneously using d_{1g} for d_{15} is another variation of the *ecological fallacy*.

In another situation, one could have group specific data for college 5 and substitute the slope estimate from these data for the population slope d_{1p}. Compare the slope from college 5 (-0.35) to d_{1p} (-0.55) from Eq. (10.7). Eq. (10.2) shows that d_{15} can be used in place of d_{1p} only when $d_{15} = d_{1k} = b_1 = d_{1p}$. This occurs only if there are no group effect b_2 and interaction effect b_3. The fallacy associated with this kind of inference has been labeled the *selective* fallacy.[2]

Another cross-level inference for the same data situation would be to infer from the slope for one college to the slope for another. For example, compare the slope from college 5 (-0.35) to the slope from college 1 (-0.61). Erroneously using one in place of another is referred to as the *contextual* fallacy (cf. Alker, 1969).

Another common situation is to have only a sample survey of the population. In this case, substituting the population slope for the value of the slope for some particular group, say college 5, is a cross-level inference based on single-source incomplete data. Eq. (10.2) shows that this inference is valid only in the absence of group and interaction effects. This form of invalid inference has been called the *universalistic* fallacy (cf. Alker, 1969).

The problem of cross-level inference also applies when moving from population or group specific data to group-level information. Suppose you really want a measure of the relationship between deviance and social norms regarding deviance for *colleges*, not individuals, but you are only in possession of a sample survey of individuals in the population. In this case, when can d_{1p} be used in place of d_{1g}? The conditions under which this can be done are the same as when we go from the group-level slope to the population slope. The invalid inference from one to the other is called the *individualistic* fallacy (cf. Alker, 1969). A related form of this fallacy is an inference from the slope for a particular group to the group-level slope.

There is another set of cross-level inference fallacies which have not been generally recognized. These involve substituting a measure of relationship from a single source of incomplete data for measures of individual or group effects. For example, this is implicit in the assumption that a slope estimated from a sample of the population is a measure of true individual effects. Compare -0.55 for the population slope to -0.68 for the estimate of the individual effect. Eq. (10.2) makes explicit the conditions under which d_{1p} would equal b_1.

[2] The labels for cross-level inference fallacies are those used by Alker (1969).

10.2 *Using Additional Information to Justify or Question Cross-level Inferences*

Cross-level inference with a single source of incomplete data basically involves substituting a known relationship for an unknown relationship. Increasing awareness of the serious risks involved led to a new line of inquiry. One may have a major data source of the "wrong" type. However, there often is additional information from other sources that might be useful. Could this additional information be used to argue that particular cross-level inferences are justifiable while others are not? A good example of this line of thinking is contained in the observation by Stokes (1969) that the regression slope obtained from aggregate election data for British constituencies was substantially different from the slope obtained from a sample survey of the electorate. From this difference Stokes inferred the existence of constituency effects.

The model represented by Eqs. (10.1) and (10.2) can be used to detect when cross-level inferences are justifiable and when they are not. Suppose one has group data and additional regarding η^2. Eq. (10.2) could be used to predict whether a slope estimated from the group data is a good estimate of the population slope. That is, the larger the value of η^2, the closer would d_{1p} be to d_{1g}. For the college data, η^2 equals 0.18. This predicts that the group-level slope is not a good predictor of the population slope.

However, this indirect reasoning is unnecessary, given the formal models of multilevel analysis and additional sources of incomplete data. The actual values of individual and context effects can be estimated. This is demonstrated subsequently. First it is useful to identify potential combinations of incomplete data.

10.3 *Data Situations Involving Combinations of Data Types*

Three basic data situations involving single-source incomplete data are identified in Table 10.2. They are group specific data, sample population data, and group data. Sometimes only one of these data types will be available. At other times two will be available. Six combinations of these data situations are examined in subsequent discussion.

The single-data types typically are obtained from different sources. Therefore an important concern is their comparability. Geographical boundaries should coincide, the variables should be conceptually equivalent, and the data should apply to the same time period. Often comparability can be verified by the data. Take the case of group data with a sample of the population. If the two data sources are comparable, the overall means obtained from each source should be close. Generally some sampling error can be expected in the population data. However, substantial discrepancies suggest that some of the assumtpions of comparability are not met.

10.4 *Introducing the Use of Combined Data Sources for Multilevel Analysis*

All the elements of a simple example of multilevel analysis with combined data sources are implicit in the preceding discussion of cross-level inference. To illustrate, imagine having comparable group data and a sample of the population for the study of getting drunk and opposition to social deviance. This means that the cells in Table 10.1 would be empty. The objective is to estimate the individual, group, and interaction effects and to fill in the missing cells.

The third equation in Eq. (10.2) can be used for this purpose. Given the combined data, all the terms in that equation, except b_1 are known. Thus,

$$\begin{aligned} d_{1p} &= b_1(1 - n^2) + b_3\bar{x}(1 - n^2) + d_{1g}n^2 \\ -0.55 &= b_1(1 - 0.18) + 0.16(0.48)(1 - 0.18) + (-0.85)(0.18) \end{aligned} \tag{10.11}$$

since $b_3 = d_{2a.1}$. Solving for b_1 gives

$$b_1 = -0.56 \tag{10.12}$$

The group effect coefficient b_2 is obtained from the multiple regression analysis of the group means, where

$$b_2 = d_{1g.2} - b_1 = -1.02 - (-0.56) = -0.46 \tag{10.13}$$

These values compare favorably to the corresponding estimates obtained from complete data.

Since $b_0 = d_{0.12} = 0.79$, the missing cell information in Table 10.1 is recovered from

$$\hat{d}_{0k} = 0.79 - 0.46\bar{x}_k \tag{10.14}$$

Table 10.3
Cell Entries Based on the Combined Data Procedure

GETTING DRUNK Y — DISAPPROVAL OF GETTING DRUNK X

College Group 1

Y	X = 0	X = 1	Total
1	557	19	576
0	210	104	314
	767	123	890

College Group 2

Y	X = 0	X = 1	Total
1	437	81	518
0	260	299	559
	697	380	1077

College Group 3

Y	X = 0	X = 1	Total
1	520	85	605
0	421	915	1336
	941	1000	1941

College Group 4

Y	X = 0	X = 1	Total
1	138	14	152
0	156	657	813
	294	671	965

College Group 5

Y	X = 0	X = 1	Total
1	19	3	22
0	31	380	411
	50	383	433

where $\hat{d}_{0k}$ are the predicted within-group intercepts. With dichotomous data, the intercepts equal the conditional proportions m_{11k}, for example, the value of m_{111} obtained from Eq. (10.14) is 0.731. Multiplying the conditional proportions by the corresponding marginal proportions gives the frequencies for that cell. The others can then be obtained through subtraction. The results of this procedure are shown in Table 10.3

10.5
Summary

This chapter begins with a summary of the formal models involved in pursuit of the two major objectives of multilevel analysis. It then addresses the question of what data are required to estimate the terms in the models. Recognizing that complete data often will not be available, the possibilities of making the estimates without complete data are explored. This leads to a simple classification of single sources of incomplete data from which one might attempt to make the necessary estimates. The pitfalls of using only one source of incomplete data are identified and labeled as the fallacies of cross-level inference. The possibilities of combining incomplete data sources are then introduced. These possibilities are examined at length in Chapters 11 and 12.

Procedures for Multilevel Analysis with Combinations of Incomplete Data

Chapter 10 concludes with the description of a specific procedure to recover missing joint distribution data in tables given a combination of group data and sample population data. We begin this chapter with a general model for the case of group data and sample population data. Procedures for other configurations of incomplete data follow. The purpose is to give formulas for estimating various parameters in cases where only partial data are available. No attempts are made to give substantive interpretations.

A single hypothetical data set with continuous variables is used. This provides the basis for comparing results from complete data to results obtained from combinations of incomplete data. Dichotomous data for 2 × 2 tables can be analyzed by these methods using dummy variables as described in Chapter 5.

11.1 Data Situation 1: Complete Data

Imagine a study of county social welfare programs involving in-home assistance to the elderly and disabled. The study is prompted by evidence of substantial amounts of variation in the hourly amount of services (homemaking chores and personal care) recommended by workers for clients having essentially the same needs. One theory is that politically liberal workers, on the average, tend to recommend more hours of service to clients than politically conservative workers. The contextual hypothesis is that workers in agencies with prevailing liberal attitudes tend to recommend more hours of service, regardless of their own political dispositions.

The study involves five counties. For this illustration, assume that there are only five counties in the state. The dependent variable Y is a weekly average of the number

Table 11.1
Data for the Hypothetical Study of Welfare Service and Political Liberalism

COUNTY NUMBER k	POLITICAL LIBERALISM SCORES x_{ik}	HOURS OF SERVICE GIVEN y_{ik}	COUNTY AVERAGES $\bar{x}_k$	$\bar{y}_k$	WITHIN-COUNTY REGRESSION LINES $y_{ik} = d_{0k} + d_{1k}x_{ik} + f_{ik}$
1	−1	1.1	1	3.32	$y_{i1} = 2.06 + 1.26x_{i1} + f_{i1}$
	0	2.4			
	1	2.1			
	2	4.8			
	3	6.2			
2	0	3.5	2	7.68	$y_{i2} = 2.94 + 2.37x_{i2} + f_{i2}$
	1	4.1			
	2	8.5			
	3	9.8			
	4	12.5			
3	1	8.6	3	12.72	$y_{i3} = 6.27 + 2.15x_{i3} + f_{i3}$
	2	10.5			
	3	12.3			
	4	15.2			
	5	17.0			
4	2	12.7	4	20.40	$y_{i4} = 6.72 + 3.42x_{i4} + f_{i4}$
	3	18.3			
	4	19.7			
	5	24.7			
	6	26.6			
5	3	21.1	5	27.52	$y_{i5} = 9.37 + 3.63x_{i5} + f_{i5}$
	4	22.9			
	5	27.4			
	6	31.0			
	7	35.2			

of service hours recommended to clients by each worker. The variable X is a measure of political liberalism based on a scale ranging from -1 to 7. Observations of X and Y are presented in Table 11.1 with agency averages on X and Y.

The population (state) level relationship between X and Y is

$$\begin{aligned} y_{ik} &= d_{0p} + d_{1p}x_{ik} + e_{ik} \\ &= 1.50 + 4.34x_{ik} + e_{ik} \end{aligned} \tag{11.1}$$

The relationship between the county averages on hours of service recommended $\bar{y}_k$ and the averages on political liberalism $\bar{x}_k$ is

$$\begin{aligned} \bar{y}_k &= d_{0g} + d_{1g}\bar{x}_k + e_k \\ &= -4.01 + 6.11\bar{x}_k + e_k \end{aligned} \tag{11.2}$$

The within-county relationships between service hours and political liberalism are analyzed in the equation

$$y_{ik} = d_{0k} + d_{1k}x_{ik} + f_{ik} \tag{11.3}$$

Values of the county slopes and intercepts from Eq. (11.3) are given in Table 11.1. The basic model of contextual analysis specifies that the intercepts and slopes are functions of the group-mean variable $\bar{X}$. The regression coefficients are shown in the equations

$$\begin{aligned} d_{0k} &= b_0 + b_2\bar{x}_k + u_k \\ &= -0.05 + 1.84\bar{x}_k + u_k \end{aligned} \tag{11.4}$$

$$\begin{aligned} d_{1k} &= b_1 + b_3\bar{x}_k + v_k \\ &= 0.83 + 0.58\bar{x}_k + v_k \end{aligned} \tag{11.5}$$

The single equation is

$$\begin{aligned} y_{ik} &= b_0 + b_1x_{ik} + b_2\bar{x}_k + b_3x_{ik}\bar{x}_k + e_{ik} \\ &= -0.05 + 0.83x_{ik} + 1.84\bar{x}_k + 0.58x_{ik}\bar{x}_k + e_{ik} \end{aligned} \tag{11.6}$$

As before, the coefficient b_1 is a measure of the individual effect of X. The coefficient b_2 is a measure of the group effect associated with the group means $\bar{X}$, and the coefficient b_3 is a measure of the cross-level interaction effect associated with X and $\bar{X}$. The values 0.83, 1.84, and 0.58 indicate the existence of a positive individual effect of political liberalism, and a positive group effect associated with county levels of liberalism. Also, there is a positive cross-level interaction effect, indicating that the relationship between service hours and political liberalism increases as county levels of political liberalism increase.

Aggregating Eq. (11.6) to the level of counties gives

$$\bar{y}_k = b_0 + (b_1 + b_2)\bar{x}_k + b_3\bar{x}_k^2 + \bar{e}_k \tag{11.7}$$

Substituting the coefficients of effects obtained from Eqs. (11.4) and (11.5) into Eq. (11.7) gives

$$\begin{aligned} \bar{y}_k &= -0.05 + (0.83 + 1.84)\bar{x}_k + 0.58\bar{x}_k^2 + \bar{e}_k \\ &= -0.05 + 2.67\bar{x}_k + 0.58\bar{x}_k^2 + \bar{e}_k \end{aligned} \tag{11.8}$$

The value of the coefficient in Eq. (11.8) associated with the squared term $\bar{x}_k^2$ suggests that a second-degree curve should be observed in the group-level relationship between the group means. A direct estimate of the group-level equation containing a squared term is

$$\begin{aligned} \bar{y}_k &= d_{0g} + d_{1g}\bar{x}_k + d_{2g}\bar{x}_k^2 + e_k \\ &= 0.07 + 2.62\bar{x}_k + 0.58\bar{x}_k^2 + e_k \end{aligned} \tag{11.9}$$

The coefficients obtained from the separate equations are close to the coefficients obtained directly from the group-level regression. This helps confirm that the model is correctly specified.

It was possible to estimate individual and context effects in this example because complete data were available. Subsequent discussion demonstrates the possibilities of making these estimates when complete data are not available.

11.2

Data Situation 2: Group Data and Sample Population Data

In this example there are group data and sample population data, but no group specific data. This could have occurred because the observations on hours of service Y and political dispositions X were obtained in two different studies. Consequently it is not known which Y goes with which X. Presumably it is impractical to get this information for a representative sample of workers for a representative sample of counties. Consequently the researcher settles for a sample of workers representative of the state.

Eq. (11.7) shows explicitly what information is missing in the group data. The regression coefficient corresponding to $\bar{x}_k$ provides the sum of the individual and group effects. With group data only, the two cannot be separated. This is where the sample population data come in.

The population slope d_{1p} is obtained from Eq. (11.1). The formula for estimating d_{1p} is derived from the normal equations for this regression. The formula is

$$d_{1p} = \frac{\sum_i \sum_k (x_{ik} - \bar{x})(y_{ik} - \bar{y})}{\sum_i \sum_k (x_{ik} - \bar{x})^2} \tag{11.10}$$

The numerator in Eq. (11.10) is the total sum of cross products of X and Y. This sum can be written as a within-group sum plus a between-group sum of cross products. This gives

$$d_{1p} = \frac{\sum_k \sum_i (x_{ik} - \bar{x}_k)(y_{ik} - \bar{y}_k) + \sum_k n_k(\bar{x}_k - \bar{x})(\bar{y}_k - \bar{y})}{\sum_i \sum_k (x_{ik} - \bar{x})^2} \tag{11.11}$$

If group specific data were available for the kth group, the slope d_{1k} would be found from Eq. (11.3). The formula for estimating d_{1k} is

$$d_{1k} = \frac{\sum_i (x_{ik} - \bar{x}_k)(y_{ik} - \bar{y}_k)}{\sum_i (x_{ik} - \bar{x}_k)^2} \tag{11.12}$$

The numerator in this expression appears as part of the numerator in Eq. (11.11). Since the numerator equals d_{1k} times the denominator, the numerator in the expression for the coefficient d_{1p} in Eq. (11.11) can be rewritten as

$$d_{1p} = \frac{\sum_k d_{1k} \sum_i (x_{ik} - \bar{x}_k)^2 + \sum_k n_k(\bar{x}_k - \bar{x})(\bar{y}_k - \bar{y})}{\sum_i \sum_k (x_{ik} - \bar{x})^2} \tag{11.13}$$

The relationship between the group coefficients d_{1k} and the effects measured by b_1 and b_3 is specified in Eq. (11.5). Substituting d_{1k} from Eq. (11.5) into Eq. (11.13), and solving for b_1, gives

$$b_1 = \frac{d_{1p} \sum_i \sum_k (x_{ik} - \bar{x})^2 - b_3 \sum_k \bar{x}_k \sum_i (x_{ik} - \bar{x}_k)^2 - \sum_k n_k(\bar{x}_k - \bar{x})(\bar{y}_k - \bar{y})}{\sum_i \sum_k (x_{ik} - \bar{x}_k)^2} \tag{11.14}$$

Concepts and procedures for explaining aggregate relationships and the data problem in multilevel analysis are discussed in Chapters 8 through 10. The intimate connection between that discussion and the present discussion of analysis with combined data sources is apparent in Eq. (11.14). The equation is essentially the same as the third equation in Eq. (10.2), showing the relationship between the aggregate slopes, the measures of effects, and the η^2. The difference is that η^2 and the group-level coefficient were made explicit in the earlier equation. It was also assumed that the within-group variances of X were relatively constant across groups. This accounts for the appearance of $\bar{x}$ in Eq. (10.2), rather than $\sum \bar{x}_k$ which appears in Eq. (11.14). The implication is that, depending on assumptions, either formula can be used to estimate b_1 in this combined data procedure.

All of the terms on the right-hand side of Eq. (11.14) can be obtained from the group data or from the sample population data. This provides an estimate of the unknown effect b_1. An estimate of b_2 can be obtained from Eq. (11.9) and the relationship $d_{1g} = b_1 + b_2$. Application to the five-county welfare problem follows.

An estimate of d_{1p} in Eq. (11.14) is obtained from the population data and Eq. (11.1). This value is 4.34. An estimate of b_3 is obtained from the group data and Eqs. (11.7) and (11.9). This value is 0.58. The other terms on the right-hand side of Eq. (11.14) are obtained from

$$\begin{aligned}
\sum_i \sum_k (x_{ik} - \bar{x})^2 &= 100 \\
\sum_i \sum_k (x_{ik} - \bar{x}_k)^2 &= 50 \\
\sum \bar{x}_k \sum (x_{ik} - \bar{x}_k)^2 &= 1(10) + 2(10) + 3(10) + 4(10) + 5(10) = 150 \\
\sum n_k(\bar{x}_k - \bar{x})(\bar{y}_k - \bar{y}) &= 5(1 - 3)(3.32 - 14.33) \\
&\quad + 5(2 - 3)(7.68 - 14.33) \\
&\quad + 5(3 - 3)(12.72 - 14.33) \\
&\quad + 5(4 - 3)(20.40 - 14.33) \\
&\quad + 5(5 - 3)(27.52 - 14.33) = 305.60
\end{aligned} \tag{11.15}$$

Substituting these estimates from the group and sample population data into Eq. (11.14) gives

$$b_1 = \frac{4.34(100) - 0.58(150) - 305.60}{50} = 0.82 \tag{11.16}$$

Given the estimate of b_1, the estimate of b_2 is obtained from

$$b_2 = d_{1g} - b_1 = 2.62 - 0.82 = 1.80$$

These results compare favorably with the results from complete data. The method provides estimates of the four coefficients b_0, b_1, b_2, and b_3 in Eqs. (11.4), (11.5), and (11.6). With these estimates it is possible to compute the within-group slopes and intercepts d_{1k} and d_{0k}. Estimates of the within, between, and total variations of X can be made from the group and population data. Together there is sufficient information to evaluate the contributions of the three effects to the explanation of worker variation in the recommendation of service hours. Discussion of the evaluation of effects is contained in Chapter 4. Also, there is sufficient information to apply the centering procedures described in Chapter 4. This procedure addresses the problem of interpretation due to multicollinearity.

In this example the estimate of the population slope d_{1p} contains no sampling error. In practice it is likely that a broad scale sample of the population would involve some sampling error, and this would affect the estimates of individual and context effects based on the incomplete data.

11.3

Data Situation 3: Group Data and Group Specific Data for One Group

There are occasions when the joint distribution for X and Y is available for a single group in addition to the marginal distributions that are available for all the groups. Such a joint distribution may consist of sample data or the complete population data for that group. For the five-county welfare problem, suppose that policies in all but the one county prohibit studies which link personal worker information to case dispositions. In this situation the group specific data for the one group can be used in several ways to estimate individual and context effects. The method proposed here gives an estimator with a smaller mean square error than other possibilities.

Estimates of the within-group intercept d_{0k} and the slope d_{1k} are obtained from Eq. (11.3). Estimates of b_0, b_3, and the sum $b_1 + b_2$ are obtained from the group data [Eq. (11.9)]. As before, the problem is to use additional data to get separate estimates for b_1 and b_2. This can be done as follows.

Eq. (11.7) specifies that the sum $b_1 + b_2$ equals the coefficient d_{1g}. For b_1 this gives

$$b_1 = d_{1g} - b_2 \quad \text{or} \quad b_1 = (b_1 + b_2) - b_2 \tag{11.17}$$

Eqs. (11.4) and (11.5) specify that the intercepts and slopes are functions of the group-mean variable $\bar{x}_k$. For the group having group specific data, the values of the intercept and slope are known. The values of b_0 and b_3 are known. The values of b_1 and b_2 are unknown. Except for the single group, the residuals u_k and v_k also are unknown.

Substituting b_1 from Eq. (11.17) into Eq. (11.5) gives

$$d_{1k} = (b_1 + b_2) - b_2 + b_3\bar{x}_k + v_k \tag{11.18}$$

To find the value of b_2, ordinary least squares methods are used to minimize the sum of squared residuals. When minimizing $u_k^2 + v_k^2$ with respect to b_2, the following estimator for b_2 is obtained.

$$\hat{b}_2 = \frac{(b_1 + b_2) - d_{1k} + (d_{0k} - b_0 + b_3)\bar{x}_k}{1 + \bar{x}_k^2} \tag{11.19}$$

The value of b_2 computed according to this expression can be used in Eq. (11.8) to find b_1. These estimators are unbiased in the usual statistical sense. Application of these procedures to the welfare example follows.

Imagine that the group specific data are for county 1. Regressing Y on X in that county gives

$$d_{01} = 2.06, \qquad d_{11} = 1.26 \tag{11.20}$$

As before, estimates of b_0, b_3, and the sum of $b_1 + b_2$ are obtained from Eq. (11.9). These estimates are

$$b_0 = 0.07, \qquad b_3 = 0.58, \qquad b_1 + b_2 = 2.62 \tag{11.21}$$

The mean of X in the first county equals 1.00. Substituting these values into Eq. (11.19) gives

$$\hat{b}_2 = \frac{2.62 - 1.26 + (2.06 - 0.072 + 0.583)(1)}{1 + 1.00^2} = 1.96 \tag{11.22}$$

This value corresponds well to the value of 1.84 obtained from complete data. Had any other of the four groups provided the group specific data, the estimates for b_2 would have been 1.43, 2.08, 1.65, and 1.86, respectively.

11.4 Data Situation 4: Group Data, Sample Population Data, and Group Specific Data for One Group

The data situation in this section is a special case where group specific data are available for several groups. As before, the regression analysis of the group means [Eq. (11.9)] provides the three estimates b_0, $(b_1 + b_2)$, and b_3. The group specific data and the sample population data are used to find the two components b_1 and b_2 of the sum $b_1 + b_2$. These two components are necessary in order to talk about individual and context effects in the explanation of worker variation in service hour recommendations. Also, they are needed to study the individual-level relationship between X and Y in the $K - 1$ tables where the group specific data are missing.

The procedure for finding b_1 and b_2 is basically the same as for complete data (Data Situation 1). Without loss of generality, it is assumed that the group specific data are available for county 1. The first step consists of using the sample population data to find the population slope d_{1p} [Eq. (11.1)]. The formula for this slope is given in Eq. (11.13). In Eq. (11.14) it is rewritten to show the presence of the within-group intercepts d_{1k}.

The difference between procedures described for that case is that the term $d_{11}\sum(x_{i1} - \bar{x}_1)^2$ is known from the group specific data for that particular county. This information can be used to find the slope d_{11}. The sum of squares for X is available from the marginal distribution for X in that county. Separate out the term for county 1 and write Eq. (11.14) as

$$d_{1p} = \frac{d_{11} \sum_i (x_{i1} - \bar{x}_1)^2 + \sum_{k>1} d_{1k} \sum_i (x_{ik} - \bar{x}_k)^2 + \sum_k n_k(\bar{x}_k - \bar{x})(\bar{y}_k - \bar{y})}{\sum_i \sum_k (x_{ik} - \bar{x})^2} \tag{11.23}$$

Substituting for d_{1k} in Eq. (11.5) and solving the resulting equation for the unknown b_1 gives an equation where everything is known except the weighted mean for $K - 1$ unknown residuals. But this mean is small and, excluding it for a moment, we get the following expression for b_1:

$$b_1' = \frac{d_{1p} \sum_k \sum_i (x_{ik} - \bar{x})^2 - d_{11} \sum_i (x_{i1} - \bar{x}_1)^2}{\sum_{k>1} \sum_i (x_{ik} - \bar{x}_k)^2} - \frac{b_3 \sum_{k>1} \bar{x}_k \sum_i (x_{ik} - \bar{x}_k)^2 - \sum_k n_k(\bar{x}_k - \bar{x})(\bar{y}_k - \bar{y})}{\sum_{k>1} \sum_i (x_{ik} - \bar{x}_k)^2} \tag{11.24}$$

The prime in Eq. (11.24) indicates that the term based on the weighted sum of the unknown residuals is excluded. The term involving the residuals is the ratio of two sums. This ratio is

$$\frac{\sum_{k>1} v_k \sum_i (x_{ik} - \bar{x}_k)^2}{\sum_{k>1} \sum_i (x_{ik} - \bar{x}_k)^2} \tag{11.25}$$

Because this mean no longer includes the first group, it cannot be expected to equal zero, and this will bias the estimate b_1. Subsequently it is shown that this bias can be adjusted. The following illustrates the application of these procedures to the welfare data.

Most of the information required in Eq. (11.24) is available from the data sources in this example. We have $d_{1p} = 4.34$. The total sum of squares for X equals 100. The slope d_{11} in the first group equals 1.25 and the sum of squares for X in that group is 10. Furthermore, b_3 equals 0.583. When county 1 is excluded, the sum $\sum \bar{x}_k \sum (x_{ik} - \bar{x}_k)^2$ equals 140. The last term in the numerator equals 305.60. When county 1 is excluded the denominator equals 40.

Substituting these values into Eq. (11.24) gives

$$b_1' = \frac{4.34(100) - 1.26(10) - 0.58(140) - 305.60}{40} = 0.85 \tag{11.26}$$

The results of this procedure correspond well to the results obtained from complete data.

Better results can be obtained by including the residuals. With the residuals included, the value of b_1 becomes

$$b_1 = b_1' - \frac{\sum_{k>1} v_k \sum_i (x_{ik} - \bar{x}_k)^2}{\sum_{k>1} \sum_i (x_{ik} - \bar{x}_k)^2} \tag{11.27}$$

Because county 1 is the only group for which there are group specific data, only the value of the residual v_1 is observed. But v_1 is not included in Eq. (11.27). To bring it in, assume that the weighted mean of the residuals is zero when all the groups are included, as follows:

$$\begin{aligned} 0 &= \sum_k v_k \sum_i (x_{ik} - \bar{x}_k)^2 \\ &= v_1 \sum_i (x_{i1} - \bar{x}_1)^2 + \sum_{k>1} v_k \sum_i (x_{ik} - \bar{x}_k)^2 \end{aligned} \tag{11.28}$$

Substituting back into Eq. (11.27) gives

$$b_1 = b_1' + \frac{v_1 \sum_i (x_{i1} - \bar{x}_1)^2}{\sum_{k>1} \sum_i (x_{ik} - \bar{x}_k)^2} \tag{11.29}$$

As a first estimate of v_1, use the model equation and solve for v_1 to get the equation

$$v_1 = d_{11} - b_1' - b_3\bar{x}_1 \tag{11.30}$$

For the welfare data this is

$$\hat{v}_1 = 1.26 - 0.85 - 0.58(1) = -0.18 \tag{11.31}$$

This gives the following adjusted value for b_1 from Eq. (11.29):

$$\hat{b}_1 = 0.85 + \frac{(-0.18)10}{40} = 0.81 \tag{11.32}$$

Repeating the process and using the new value of b_1, the residual becomes

$$\hat{v}_1 = 1.26 - 0.81 - 0.58(1) = -0.13 \tag{11.33}$$

This leads to the following value of b_1:

$$\hat{b}_1 = 0.85 + \frac{(-0.13)10}{40} = 0.82 \tag{11.34}$$

By repeating this process several more times the value of b_1 settles down at 0.82. The adjustment of b_1 has an effect here because there are only five groups, making it possible for a single residual to have an effect on the initial value of b_1. With larger numbers of groups such an adjustment as seen here should not be necessary.

There is a difference between the way b_1 is found here and in Data Situation 2, where only sample population data are available. Here the slope for the first group is explicitly introduced, and one does not have to depend on the basic formal model for that group in the initial computation. Instead, the model enters when the adjustment is performed for the residual term.

11.5 *Data Situation 5: Group Data and Group Specific Data for Several Groups*

Having group specific data available for several groups is an intermediate situation between Data Situation 3, where the group specific data are available for only one group, and Data Situation 1, where the group specific data are available for all the groups. In this intermediate situation there is a choice of methods for computing the four b's in the two separate model equations. The choice centers mainly around the question of how many groups have individual-level data. If such data are available for almost all the groups, it would be better to use the methods described under Data Situation 1. On the other hand, with group specific data available for only some groups, one should base the estimation on the regression analysis of the group means and supplement this by the information contained in the group specific data. Ways to accomplish this are described in this section. It is difficult to develop firm criteria for which method is best in any particular circumstance, and the choice often becomes a matter of judgment.

In this section assume there are group specific data available for some of the groups. The group means for all K groups are used to find the three estimates b_0, $(b_1 + b_2)$, and b_3. Use the additional group specific data available for K' of the groups to obtain the separate estimates of b_1 and b_2. Here K' is some number less than the number of groups (counties).

For each of the K' groups where the individual-level data are known, use those data to find the intercept d_{0k} and the slope d_{1k}. According to the basic model, these quantities are related to the b's as seen in Eqs. (11.4) and (11.5). From the regression analysis of the group means, values of b_0 and b_3 are known, as in the sum $b_1 + b_2 = d_{1g}$ in Eq. (11.9). The separate components b_1 and b_2 are not known. If b_2 were known, b_1 could be found according to the equation

$$b_1 = (b_1 + b_2) - b_2 = d_{1g} - b_2 \tag{11.35}$$

Substitute this expression for b_1 back into Eq. (11.5) to get

$$d_{1k} = (b_1 + b_2) - b_2 + b_3\bar{x}_k + v_k \tag{11.36}$$

There is a total of $2K'$ equations in Eq. (11.36) with one unknown (b_2) and a collection of residuals. One way to find a value of b_2 is to use ordinary least squares methods to minimize the sum of squares of the residuals with respect to b_2. That is, minimize the sum

$$\begin{aligned} Q &= \textstyle\sum'(u_k^2 + v_k^2) \\ &= \textstyle\sum'[(d_{0k} - b_0 - b_2\bar{x}_k)^2 + (d_{1k} - (b_1 + b_2) + b_2 - b_3\bar{x}_k)^2] \end{aligned} \tag{11.37}$$

with respect to b_2. The prime on the summation sign indicates that we are only adding across the K' groups for which the group specific data are given. The resulting expression for b_2 is

$$\hat{b}_2 = \frac{K'(b_1 + b_2) - \sum' d_{1k} + \sum' d_{0k}\bar{x}_k - (b_0 - b_3)\sum'\bar{x}_k}{K' + \sum'\bar{x}_k^2} \tag{11.38}$$

If the number of observations differ widely in the groups, the differences should be considered explicitly in computing b_2. Let the number of individuals in the kth group be equal to n_k. The formula for b_2 is

$$\hat{b}_2 = \frac{(b_1 + b_2)\sum' n_k - \sum' n_k d_{1k} + \sum' n_k d_{0k}\bar{x}_k - (b_0 - b_3)\sum' n_k \bar{x}_k}{\sum' n_k + \sum' n_k \bar{x}_k^2} \tag{11.39}$$

Finally, with the value of b_2 from Eq. (11.39) we can find b_1 from Eq. (11.35).

The formula in Eq. (11.38) is a direct generalization of the similar formula in Data Situation 3. Applying the formula to the welfare data, we get a range of values for b_2 from 1.43 to 2.08 when the group specific data are available for one group. When group specific data are available for two counties ($K' = 2$) and b_2 is found for all combinations of two counties, the range of values for b_2 narrows to 1.58 to 2.06. As an example, consider the value obtained for b_2 when the group specific data are available for counties 1 and 4. From Table 11.1 we have $d_{01} = 2.06$ and $d_{04} = 6.72$, $d_{11} = 1.26$ and $d_{14} = 3.42$, $\bar{x}_1 = 1$, and $\bar{x}_4 = 4$. Using Eq. (11.38) gives

$$\hat{b}_2 = \frac{2(2.615) - 4.68 + 28.94 - (-0.511)5}{2 + 17} = 1.69 \tag{11.40}$$

Using all combinations of three groups, the values for b_2 range from 1.63 to 1.92; with four groups the range is from 1.75 to 1.87, and with all five groups we get the single value of 1.81.

11.6

Data Situation 6: Group Data, Sample Population Data, and Group Specific Data for Several Groups

Now consider the case where there are group specific data for several but not all the groups, and a sample survey of the population. It makes some difference whether the individual-level data are available for only a few of the groups or whether they are available for most of the groups. If the group specific data are available for most of the groups, then there is not much additional information contained in the sample population data. Most of the population data can be obtained simply by adding up the data already present in the separate tables. When this is the case, return to the procedures for Data Situation 5 and use that procedure to estimate the individual and context effects, and for recovering the missing group specific data.

If the group specific data are not available for most of the groups, it makes sense to utilize the additional information contained in the population sample. The procedure developed in this section for estimating the four model parameters follows directly from preceding sections. The estimates of b_0, $(b_1 + b_2)$, and b_3 are obtained from the regression of the group means. Using the method for Data Situation 5, separate estimates of b_1' and b_2' are obtained from the groups containing group specific data. The value of b_1' is used to generate a set of residuals that are then used to find a new value b_1'' according to the ideas presented for Data Situation 2. Finally, the two values of b_1 are averaged in a weighting procedure explained at the end of this section.

In this data situation we have marginal distributions for the two variables X and Y in each of K groups and for K' of these groups the joint distribution of X and Y. In addition we have the joint distribution of X and Y in the population comprised of the individuals belonging to these groups. One way to find the values of the four b's in the basic model is as follows.

First the means of the marginal distributions can be used to find estimates of b_0, $(b_1 + b_2)$, and b_3. For the welfare data these quantities are 0.07, 2.62, and 0.58. Second, the group specific data in the K' groups are used to find the separate estimates b_1' and b_2'. Suppose K' equals 2 and that we have the group specific data for counties 1 and 4. In the previous section it was found that these two counties lead to the estimates $b_1' = 2.62 - 1.69 = 0.93$ and $b_2' = 1.69$. The data in the sum table can now be used to improve on these two estimates.

Eq. (11.13) gives a formula for the population slope d_{1p}. The first term in the numerator for that formula can now be written

$$\sum_k^K d_{1k} \sum_i (x_{ik} - \bar{x}_k)^2 = \sum^{K'} d_{1k} \sum (x_{ik} - \bar{x}_k)^2 + \sum^{K-K'} d_{1k} \sum (x_{ik} - \bar{x}_k)^2 \tag{11.41}$$

The first summation on the right-hand side of the equation represents the sum over the K' groups where the group specific data are known and where the slopes d_{1k} therefore are known. The second sum is the sum across the remaining $K - K'$ groups where the group specific data are not known. For these groups substitute for d_{1k} from the basic model according to the equation

$$d_{1k} = b_1 + b_3 \bar{x}_k + v_k \tag{11.42}$$

With this substitution, Eq. (11.41) can be changed to read

$$\begin{aligned} \sum d_{1k} \sum (x_{ik} - \bar{x}_k)^2 &= \sum^{K'} d_{1k} \sum (x_{ik} - \bar{x}_k)^2 + \sum^{K-K'} (b_1 + b_3 \bar{x}_k + v_k) \sum (x_{ik} - \bar{x}_k)^2 \\ &= \sum^{K'} d_{1k} \sum (x_{ik} - \bar{x}_k)^2 + b_1 \sum^{K-K'} \sum (x_{ik} - \bar{x}_k)^2 \\ &\quad + b_3 \sum^{K-K'} \bar{x}_k \sum (x_{ik} - \bar{x}_k)^2 + \sum^{K-K'} v_k \sum (x_{ik} - \bar{x}_k)^2 \end{aligned} \tag{11.43}$$

The residual terms and the coefficient b_1 are not known in Eq. (11.43). But it is possible to get an estimate of the sum involving the residuals from the values we already have. The sum contains the residuals for the $K - K'$ groups where the group specific data are not known. But if we assume that the sum of the residuals across all the groups equals zero, then

$$\sum^{K'} v_k \sum (x_{ik} - \bar{x}_k)^2 + \sum^{K-K'} v_k \sum (x_{ik} - \bar{x}_k)^2 = 0 \tag{11.44}$$

Thus the sum where the residuals are unknown can be replaced with the sum where the residuals are known and the sign changed. For the welfare data this gives

$$\begin{aligned} v_1 &= d_{11} - b_1' - b_3 \bar{x}_1 \\ &= 1.26 - 0.93 - 0.583(1) = -0.253 \\ v_4 &= 3.42 - 0.93 - 0.583(4) = 0.158 \end{aligned} \tag{11.45}$$

$$\sum^{K'} v_k \sum (x_{ik} - \bar{x})^2 = -0.253(10) + 0.158(10) = -0.95$$

By taking the expression in Eq. (11.43) and substituting it back into Eq. (11.13), we can solve for the new value of b_1 and get

$$\begin{aligned} b_1'' = \Bigg[& d_{1p} \sum^{K} \sum (x_{ik} - \bar{x})^2 - \sum^{K'} d_{1k} \sum (x_{ik} - \bar{x}_k)^2 - b_3 \sum^{K-K'} \bar{x}_k \sum (x_{ik} - \bar{x}_k)^2 \\ & + \sum^{K'} v_k \sum (x_{ik} - \bar{x}_k)^2 - \sum^{K} n_k (\bar{x}_k - \bar{x})(\bar{y}_k - \bar{y}) \Bigg] \Bigg/ \sum^{K-K'} \sum (x_{ik} - \bar{x}_k)^2 \end{aligned} \tag{11.46}$$

The expression for b_1 is based on the group specific data in the K' groups as well as the sample population data. For the welfare data,

$$\begin{aligned} b_1'' = [& 4.34(100) - (1.26 + 3.42)10 - 0.58(2(10) + 3(10) + 5(10)) - 0.95 \\ & - 305.60]/(10 + 10 + 10) = 0.75 \end{aligned} \tag{11.47}$$

The value for b_1 obtained from complete data was 0.83. Here we started with a value of 0.93 from the group specific data in counties 1 and 4. When the sample population data were introduced, the value for b_1 changed to 0.74, which is closer to 0.83 than the first value.

If the value of b_1 obtained from the K' tables with known group specific data and used to find the residuals is too large, it is now possible to show that the value of b_1 from Eq. (11.46) is too small, and vice versa. To compensate for this, use a weighted mean of the two values as the final value of b_1. Here b_1' is the value obtained from the group specific data in the K' counties according to the methods for Data Situation 5, and b_1'' is the value from Eq. (11.46). Then the weighted mean b_1 is found from the expression

$$\hat{b}_1 = \frac{b_1' \sum^{K'} \sum (x_{ik} - \bar{x}_k)^2 + b_1'' \sum^{K-K'} \sum (x_{ik} - \bar{x}_k)^2}{\sum^{K} \sum (x_{ik} - \bar{x}_k)^2} \tag{11.48}$$

The weight for b'_1 is the sum of the sums of squares for X in the K' groups where the group specific data are not known.

The value for the welfare data is

$$b_1 = \frac{0.93(20) + 0.74(30)}{50} = 0.82 \tag{11.49}$$

Using other pairs of counties and the population sample gives values close to 0.82 obtained from complete data.

11.7
Multilevel Analysis with Incomplete Data, Using Correlations as Inputs to Regression Analyses

Many statistical packages for computer analysis allow the input of observed correlation coefficients into regression instead of unsummarized observations for each individual case. This suggests a strategy for multilevel analysis when complete data are not available. In brief, the strategy is to obtain estimates of means, standard deviations, and correlation coefficients from whatever partial data sources are available in order to obtain the partial regression coefficients specified in single-equation contextual analyses. For example, the correlation coefficient relating X and Y for every case in a population cannot be directly obtained without complete data. However, it can be estimated given a sample of the population. In combination with group data, each of the regression coefficients in the model can be estimated. This alternative procedure is somewhat more general in application than the procedures requiring specific formulas, as above. Details for this procedure are contained in Appendix I.

11.8
Summary

A major obstacle to multilevel analysis is the amount of data required. This chapter demonstrates that it is theoretically possible to carry out multilevel analyses when complete data are not available. Specific formulas are given for estimating the parameters in contextual analyses when various configurations of partial data are available. Substantive examples are given in the next chapter.

A somewhat more general approach also is introduced. This approach involves the use of information from various partial data sources which is summarized in the form of means, standard deviations, and correlation coefficients. Details for this approach are contained in Appendix I.

Additional Examples of Multilevel Analysis with Incomplete Data

This chapter presents two examples of multilevel analysis with incomplete data. The first example is based on a study of political change in Britain. This study is introduced in Chapter 2. It illustrates the case where group data are supplemented by a sample of the population to explain a puzzling group-level phenomenon. The second example refers to the study of illiteracy and nativity examined in Chapter 9. It illustrates the application of procedures described in Chapter 11 to explain complex nonlinear curves at the group level.

12.1 *The Puzzle of Uniform Election Swings in Britain*

Butler and Stokes (1969) provide an early example of multilevel analysis with incomplete data in their study of political change in Britain. Their analysis involved "the puzzle of uniform swings." Table 12.1 shows a table with marginal proportions for a single constituency for the 1964 and 1966 elections. In this constituency the Conservative party received the proportion $p_{\cdot 2k}$ of the vote in 1964 and the proportion $p_{1 \cdot k}$ of the vote in 1966. The difference between these two proportions is defined as the swing for this constituency. It was found that swings varied little from constituency to constituency. The objective of this analysis is to explain why.

The phenomenon of constant swings is formally represented by

$$\text{swing} = p_{1 \cdot k} - p_{\cdot 2k} \approx \text{constant} \qquad (12.1)$$

With the swing data alone, the cell entries for each constituency are unknown. Without further information it is impossible to study which individuals voted the same in both elections and which individuals changed their votes.

Table 12.1
Turnover Table with Marginal Proportions and Missing Cell Entries for the kth Constituency for the 1964 and 1966 British Elections Together with Values for the Two Dummy Variables Representing the Elections

		Vote in 1964 Election X		
		0 Not conservative	1 Conservative	Total
Vote in 1966 Election Y	1 Conservative			$p_{1 \cdot k}$
	0 Not conservative			$p_{2 \cdot k}$
	Total	$p_{\cdot 1k}$	$p_{\cdot 2k}$	1.00

A popular interpretation was that national forces explained the observed uniformity in election swings. Taking issue with this popular interpretation, Butler and Stokes argued that the phenomenon was due to a special combination of national *and* local forces which tended to cancel out at the level of constituencies. Their argument was based on the results of a sample survey of the whole electorate. The issue is explored here by identifying the implications of several competing theories and checking them against the group data and the sample population data.

12.1.1. Theories of Individual Voting Behavior

Perhaps the simplest theory about the missing cell entries in Table 12.1 is to assume that voting habits or party attachment completely dictates a voter's choice of candidate. In that case the upper left and lower right cells of Table 12.1 would be empty and all the voters would be found in the other two cells. No one would change their vote from election to election, and elections would be won and lost solely on the basis of turnout and physical changes in the electorate. A direct consequence of this theory is that for a given constituency the proportion voting for a particular party at one election equals the proportion voting for the same party at the previous election, and the swing would equal zero. The idea is unrealistic and unsupported by the data. When the cell entries based on the theory of perfect party attachment are aggregated to the level of the population, the conditional proportions equal 1 and 0. Examination of the conditional proportions in Table 12.2 clearly eliminates the hypothesis of perfect party attachment.

The survey data in Table 12.2 suggest that there are considerable deviations from perfect party allegiance. One explanation for these deviations is that national political issues and events induce individuals to change their votes from one election to the next, and these forces are felt in the same way in all constituencies. This implies that the conditional proportions m_{11k} and m_{12k} do not vary from constituency to constituency. They could therefore be obtained from the sample population data. This theory can be tested by entering these values and computing estimates of the 1966

Table 12.2
Vote Proportions from Sample Survey Data with Conditional Proportions in Parentheses

		Vote in 1964 Election		
		Not conservative	Conservative	Total
Vote in 1966 Election	Conservative	0.05 (0.08)	0.26 (0.80)	0.31
	Not conservative	0.62 (0.92)	0.07 (0.20)	0.69
	Total	0.67 (1.00)	0.33 (1.00)	1.00

marginal proportions from the 1964 marginal proportions, and by comparing the estimated 1966 proportions to the observed proportions for that year. Such a comparison eliminates the theory of national effects only.

The third theory is that voter deviation from perfect party attachment varies systematically across constituencies with the relative strength of the parties. This implies that the column proportions m_{11k} and m_{12k} are greater than the national-level proportions in strong Conservative constituencies, and smaller in weak Conservative constituencies. Since the estimated proportions are found to follow such a pattern, it can be said that the puzzle of election swings is explained by this theory of individual voting behavior.

12.1.2. Application of Multilevel Analysis Procedures

The above analysis is made explicit by applying the formal multilevel analysis procedures described in Chapter 11. To talk about individual-level voting behavior, we need the missing conditional column proportions m_{11k} and m_{12k}. The proportion of people in the kth constituency voting Conservative in 1966 and voting not Conservative in 1964 is m_{12k}. Similarly, m_{12k} is the proportion voting Conservative in 1966 of those voting Conservative in 1964. These proportions can be expressed as functions of the marginal proportions $p_{\cdot 2k}$ as in

$$\begin{aligned} m_{11k} &= b_0 + b_2 p_{\cdot 2k} + u_k \\ m_{12k} &= (b_0 + b_1) + (b_2 + b_3) p_{\cdot 2k} + (u_k + v_k) \end{aligned} \tag{12.2}$$

With marginal data alone, the coefficients in Eq. (12.2) cannot be obtained. However, it is known that the marginal proportions are related according to the equation

$$p_{1\cdot k} = b_0 + (b_1 + b_2) p_{\cdot 2k} + b_3 p_{\cdot 2k}^2 + \bar{e}_k \tag{12.3}$$

Because the uniformity of elections has been observed, we know something about this equation. For the difference $p_{1 \cdot k} - p_{\cdot 2k}$ to be approximately constant across constituencies, the coefficients for the variables in Eq. (12.3) should be

$$b_1 + b_2 = 1.00, \qquad b_3 = 0.00 \tag{12.4}$$

Given these values, the swings are approximately constant. They can be expressed in

$$p_{1 \cdot k} - p_{\cdot 2k} = b_0 + \bar{e}_k \tag{12.5}$$

Thus b_0 is the value of the average swing. When $p_{1 \cdot k}$ is regressed on $p_{\cdot 2k}$ and $p^2_{\cdot 2k}$, the values of the coefficients in Eq. (12.3) are obtained. The constant swing has the value

$$b_0 = -0.02 \tag{12.6}$$

With b_3 equal to zero, there is no interaction effect. With no interaction effect, the swing is constant because the coefficients for the individual and group variables add up to one.

A better understanding of the individual and group effects requires separating the sum $(b_1 + b_2)$ into the two components b_1 and b_2. This is possible when a sample of the population is available. Estimates obtained from Table 12.2 are

$$b_1 = 0.69, \qquad b_2 = 0.31 \tag{12.7}$$

The values of b_1 and b_2 are substituted into Eq. (12.2) to get

$$\begin{aligned} m_{11k} &= -0.02 + 0.31p_{\cdot 2k} + u_k \\ m_{12k} &= 0.67 + 0.31p_{\cdot 2k} + (u_k + v_k) \end{aligned} \tag{12.8}$$

Disregarding the residual terms in Eq. (12.8), the conditional proportions provide estimates of the missing cell entries in each constituency. Table 12.3 shows the estimated conditional proportions in constituencies classified as low, medium, and high on 1964 proportions of Conservative votes.

12.1.3. Interpretation

The estimates in Table 12.3 suggest that the Conservatives lost ground in 1966 because they did not hold their 1964 votes as well as the other parties did. The first table shows that, in a constituency where there was a 0.75 Conservative vote in 1964, 90 percent of the Conservative voters voted Conservative again. But the last table shows that, when there was a 0.75 non-Conservative vote in 1964, as much as 94 percent of the non-Conservatives voted non-Conservative again in the next election. Similarly,

Table 12.3
Estimated Column Conditional Proportions for Three Constituencies with Given Margins for 1964 and Estimated Margins for 1966

HIGH PROPORTIONS CONSERVATIVE

		1964 Not conservative	1964 Conservative	Margin
	Conservative	0.21	0.90	0.73
1966	Not conservative	0.79	0.10	0.27
	Margin	0.25	0.75	1.00

MEDIUM PROPORTIONS CONSERVATIVE

		1964 Not conservative	1964 Conservative	Margin
	Conservative	0.14	0.83	0.48
1966	Not conservative	0.86	0.17	0.52
	Margin	0.50	0.50	1.00

LOW PROPORTIONS CONSERVATIVE

		1964 Not conservative	1964 Conservative	Margin
	Conservative	0.06	0.75	0.23
1966	Not conservative	0.94	0.25	0.77
	Margin	0.75	0.25	1.00

in the middle table where the vote was evenly split in 1964, 83 percent of the 1964 Conservative voters did not change their votes while 86 percent of the 1964 non-Conservative voters did not change their votes at the next election. Thus the 1964 Conservative voters display more change in votes than the non-Conservative voters.

The group effect is measured by the coefficient b_2, which is equal to 0.31. This coefficient measures the local effect of the composition of the constituency on the 1966 vote. The effect shows up in the increases in the conditional proportions voting Conservative in 1966 as the marginal proportion voting Conservative in 1964 increases. If we consider the proportions as probabilities, it is possible to give the local effect the following probabilistic interpretation. When an individual moves from a constituency where nobody voted Conservative in 1964 ($p_{\cdot 2k} = 0$) to a constituency where everybody voted Conservative in 1964 ($p_{\cdot 2k} = 1$), the probability of this individual voting Conservative in 1966 increases by 0.31. The value of this increase is given by the coefficient for the group variable, when there is no interaction effect in the data.

Comparing those who voted Conservative with those who did not vote Conservative in 1964, the data for the three constituencies in Table 12.3 also show that the difference between the proportions voting Conservative in 1966 for these two categories equals 0.69 in all three constituencies. This difference, $m_{12k} - m_{11k}$, equals the slope d_{1k}, and these slopes are equal because there is no interaction effect. In terms of probabilities, someone voting Conservative in 1964 had a probability of voting Conservative in 1966 that is 0.69 higher than the probability for someone who did not vote Conservative in 1964. The difference is the effect on the vote by the individual characteristic of voting Conservative in 1964. This effect is measured by the coefficient $b_1 = 0.69$, which is the coefficient for the individual-level variable.

The uniform swing in the vote proportions occurs because the sum of the two coefficients b_1 and b_2 equals 1. To better understand the meaning of this sum, consider the quantity $1 - b_1$. Imagine for a moment that the vote in 1964 completely determines how an individual votes in 1966. Given no changes in the votes from 1964 to 1966, all votes would appear on the main diagonal of Table 12.3; the conditional proportions m_{11k} would equal 0.0, and the conditional proportions m_{12k} would equal 1.0. The 1964 individual-level characteristic completely determines the 1966 vote, and the coefficient for the individual-level variables equals one, that is,

$$b_1 = d_{1k} = m_{12k} - m_{11k} = 1.00 - 0.00 = 1.00 \tag{12.9}$$

However, some people did change their votes in the two elections, bringing the value of b_1 down from 1.00. We attribute this turnover in votes to a national force affecting all constituencies in the same way. This national effect can be measured as the difference between the extreme value of b_1 in Eq. (12.9) and the observed value of b_1 equal to 0.69. Let b_N be the coefficient for this national force, where b_N becomes

$$b_N = 1 - b_1 = 1.00 - 0.69 = 0.31 \tag{12.10}$$

There is another way to look at the amount of turnover in the votes between the two elections. The sum of the two conditional proportions m_{11k} and m_{22k} gives the amount of changes in the votes, while controlling for the unequal margins for the 1964 election. This is because m_{11k} is the proportion of the 1964 non-Conservative voters who vote Conservative in 1966 and m_{22k} is the proportion of the 1964 Conservative voters who vote non-Conservative in 1966. Add these two proportions and leave out the residuals, as follows:

$$\begin{aligned} m_{11k} + m_{22k} &= m_{11k} + (1 - m_{12k}) \\ &= (-0.02 + 0.31p_{.2k}) + (1 - 0.67 - 0.31p_{.2k}) = 0.31 \end{aligned} \tag{12.11}$$

This sum is the same as b_N in Eq. (12.10).

A uniform swing in the marginal proportions is observed because b_N, the coefficient for the national force producing vote changes, has the same value as b_2, the coefficient for the local force. This can be observed in Eq. (12.3). Subtracting

$p_{\cdot 2k}$ and recalling that b_3 equals zero, the swing equals

$$\begin{aligned} \text{swing} &= p_{1 \cdot k} - p_{\cdot 2k} \\ &= b_0 + (b_2 - (1 - b_1))p_{\cdot 2k} + \bar{e}_k \\ &= b_0 + (b_2 - b_N)p_{\cdot 2k} + \bar{e}_k \end{aligned} \tag{12.12}$$

Thus the swing is constant and equal to b_0 when b_2 equals b_N. National and local effects cancel out, producing a constant swing in all constituencies and explaining the puzzle of uniform election swings.

12.2 *Explaining a Complex Curve in the Study of Illiteracy and Nativity with Incomplete Data*

This illustration applies to the problem of illiteracy and nativity introduced in Chapter 9 (cf. Robinson, 1950). The data and notation are the same. The task is to explain the complex curve relating state proportions of illiteracy and foreign-borns. This curve is shown in Figure 9.4.

The data can be visualized in 2 × 2 tables reflecting 48 within-state relationships between illiteracy and nativity. To observe the difficulties of working with group data alone, imagine that the table margins are known and the table cells are unknown. The missing cell entries can be inferred from the group data with the assumption of an individual effect only model of behavior. That model implies that within-group intercepts and slopes are constant. Under this assumption, these constants equal the regression coefficients d_{0g} and d_{1g} in the bivariate group-level regression equation. These values are

$$\begin{aligned} \bar{y}_k &= d_{0g} + d_{1g}\bar{x}_k + e_k \\ &= 0.07 - 0.17\bar{x}_k + e_k \end{aligned} \tag{12.13}$$

Because it is assumed that this is a case of individual effect only, the intercept and slope in each group equal the intercept and slope for the group means d_{0g} and d_{1g}. Consequently, the conditional proportions m_{fk} and m_{nk} are obtained from the relationships

$$\begin{aligned} m_{fk} &= d_{0k} + d_{1k} = d_{0g} + d_{1g} \\ &= 0.07 - 0.17 = -0.10 \end{aligned} \tag{12.14}$$

$$m_{nk} = d_{0k} = d_{0g} = 0.07 \tag{12.15}$$

The implications of this procedure are presented in Figure 12.1. Straight dashed lines represent the values m_{fk} and m_{nk} inferred from Eqs. (12.14) and (12.15). The

Figure 12.1
Comparison of Estimates for Illiteracy Rates Based on Group Data Alone to Those Actually Predicted with Complete Data

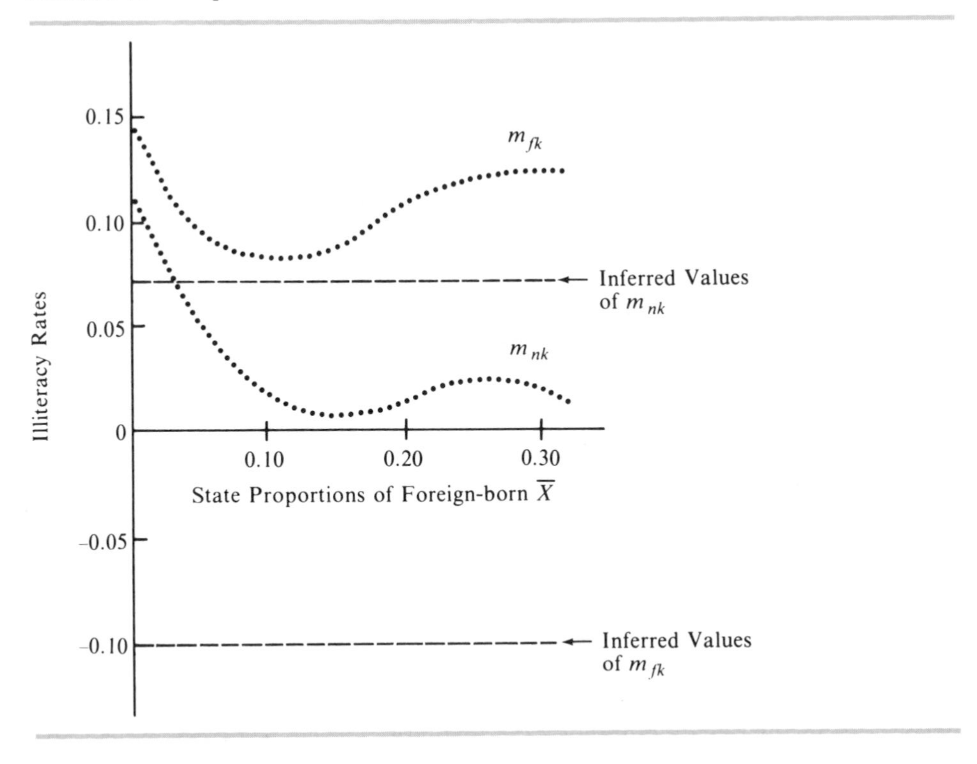

dotted curves (obtained from Figure 9.5) represent the actually predicted observations. The relationships between illiteracy and nativity inferred from the group data are exactly opposite to reality. Moreover, the estimates are inadmissible because conditional proportions cannot have values smaller than zero.

To illustrate given another incomplete data situation, imagine that the only data available are from a sample survey of the whole country. The annual Current Population Studies of the Census Bureau produces data of this type. In this case, the individual effect only model of behavior implies that the within-group intercepts d_{0k} are constant and equal d_{0p}, and that the slopes d_{1k} are constant and equal d_{1p}. Consequently, the values of the conditional proportions are inferred from

$$m_{fk} = d_{0k} + d_{1k} = d_{0p} + d_{1p}$$
$$= 0.04 + 0.06 = 0.10 \tag{12.16}$$

$$m_{nk} = d_{0k} = d_{0p} = 0.04 \tag{12.17}$$

The inferred values are compared to the actual values in Figure 12.2. The inferred within-state relationships are portrayed correctly with respect to sign. However, the

Figure 12.2
Comparison of Estimates for Illiteracy Rates Based on Sample Population Data Alone to Those Actually Predicted with Complete Data

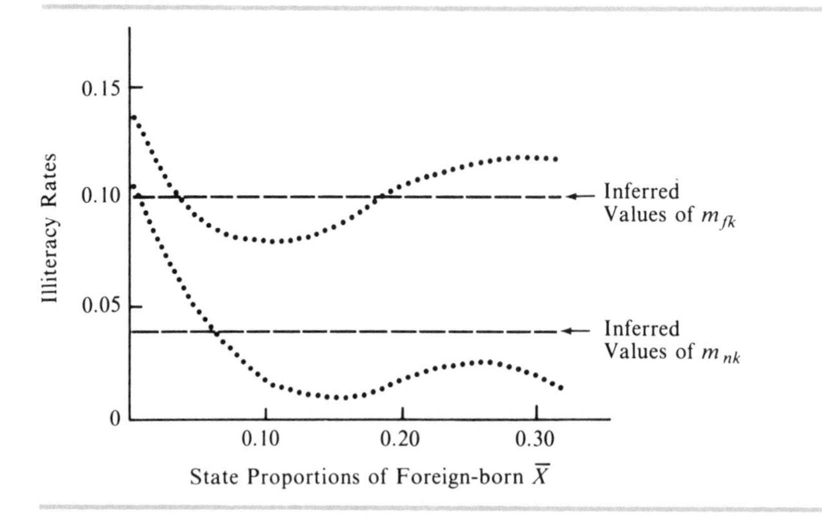

illiteracy rates are substantially overestimated at some levels of the mean variable $\bar{x}_k$, and substantially underestimated at other levels.

It has been repeatedly observed that analyses considering only individual effects generally provide inadequate explanations of group-level relationships. The basic model of contextual analysis specifies that within-group intercepts and slopes are functions of the group-mean variable. Since the observed group-level relationship is a complex polynomial curve, nonlinear context effects can be expected in the separate equations. The separate equations for the model containing terms for second- and third-degree group effects and a second-degree term for interaction effects are specified in Eqs. (9.49) and (9.50). The single equation is given in Eq. (9.51). The group-level equation obtained by aggregating the single equation is given in Eq. (9.52). Comparing that equation to the group-level multiple-regression equation [Eq. (9.48)] pinpoints the difficulty of doing multilevel analysis with group data only. There are seven coefficients to be estimated from three independent variables.

With group data only, the intercept and the sums of the various pairs of coefficients can be estimated. These estimates are

$$\begin{aligned} b_0 &= 0.11 \\ (b_1 + b_2) &= -1.79 \\ (b_3 + b_4) &= 10.78 \\ (b_5 + b_6) &= -18.60 \end{aligned} \tag{12.18}$$

As in previous examples, additional information in the form of a sample of the population can help separate pairs of coefficients observed as single values at the group level. The formal model of multilevel structure relating the population slope d_{1p} to the other coefficients and η^2 is

$$d_{1p} = b_1(1 - \eta^2) + b_3\bar{x}(1 - \eta^2) + b_5\overline{\bar{x}^2}(1 - \eta^2) + d_{1g}\eta_2 \quad (12.19)$$

where $\overline{\bar{x}^2}$ is the weighted mean of the squared mean of x_k^2, using η_k as weights, and d_{1g} is the coefficient for $\bar{x}_k$ in the linear relationship between the group means. Given group data and sample population data, there are three unknowns, b_1, b_3, and b_5. This indicates that two must be assumed to equal zero in order to get a value for the third. The resulting estimates then can be substituted into Eq. (12.18) to get the values of b_2, b_4, and b_6.

Ideally, deciding which two of the six coefficients in Eq. (12.18) to set to zero ought to be based on theory or prior knowledge of the problem. One possibility is that there is no relationship between illiteracy and nativity after controlling for all state context effects. This amounts to setting b_1 equal to zero. Another possibility is that there are no second-degree interaction effects. This sets b_5 equal to zero.

An estimate of b_3 from Eq. (12.19) becomes

$$\begin{aligned}\hat{b}_3 &= (d_{1p} - d_{1g}\eta^2)/\bar{x}(1 - \eta^2)\\ &= (0.06 - (-0.17)0.12)/0.14(1 - 0.12) = 0.65\end{aligned} \quad (12.20)$$

An estimate of b_4 obtained from Eq. (12.18) is

$$\hat{b}_4 = (b_3 + b_4) - \hat{b}_3 = 10.78 - 0.65 = 10.13 \quad (12.21)$$

Thus the estimates for the coefficients in the separate equations are

$$\begin{aligned}d_{0k} &= b_0 + \hat{b}_2\bar{x}_k + \hat{b}_4\bar{x}_k^2 + \hat{b}_6\bar{x}_k^3 + u_k\\ &= 0.11 - 1.79\bar{x}_k + 10.13\bar{x}_k^2 - 18.60\bar{x}_k^3 + u_k\end{aligned} \quad (12.22)$$

$$\begin{aligned}d_{1k} &= \hat{b}_1 + \hat{b}_3\bar{x}_k + \hat{b}_5\bar{x}_k^2 + v_k\\ &= 0.00 + 0.65\bar{x}_k + 0.00\bar{x}_k^2 + v_k\end{aligned} \quad (12.23)$$

Adding these equations as before to get the relationships involving the conditional proportions gives

$$\hat{m}_{nk} = 0.11 - 1.79\bar{x}_k + 10.13\bar{x}_k^2 - 18.60\bar{x}_k^3 + u_k \quad (12.24)$$

$$\hat{m}_{fk} = 0.11 - 1.14\bar{x}_k + 10.13\bar{x}_k^2 - 18.60\bar{x}_k^3 + u_k + v_k \quad (12.25)$$

The results shown in Figure 12.3 provide a comparison to the actual curves based on complete data. This demonstrates that it is possible to closely approximate complex group-level curves when in possession of group data and a sample of the population. However, different *a priori* decisions about which coefficients to set to zero could produce different results.

Figure 12.3
Comparison of Estimates for Illiteracy Rates Based on the Combined-data Procedure to Those Actually Predicted with Complete Data

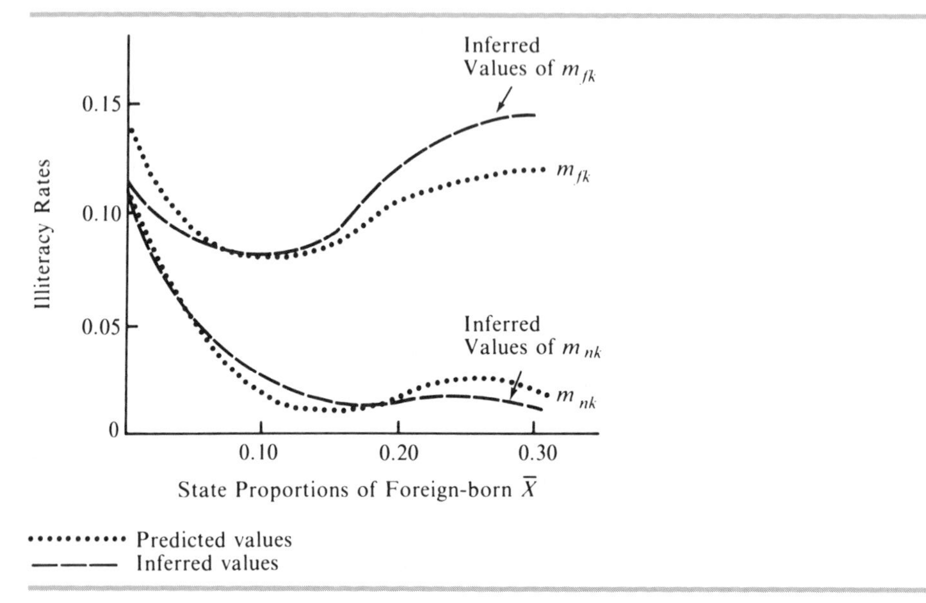

12.3
Summary

Chapter 11 presents a technical discussion of formulas used to carry out multilevel data analyses with various combinations of incomplete data. This chapter provides two substantive examples of the application of the techniques. The first example is based on a study of political change in Britain, illustrating the case where group data are combined with a sample of the population to explain a puzzling group-level phenomenon. The second example examines the study of illiteracy and nativity introduced in Chapter 9 to illustrate the combination of data sources to explain complex group-level phenomena.

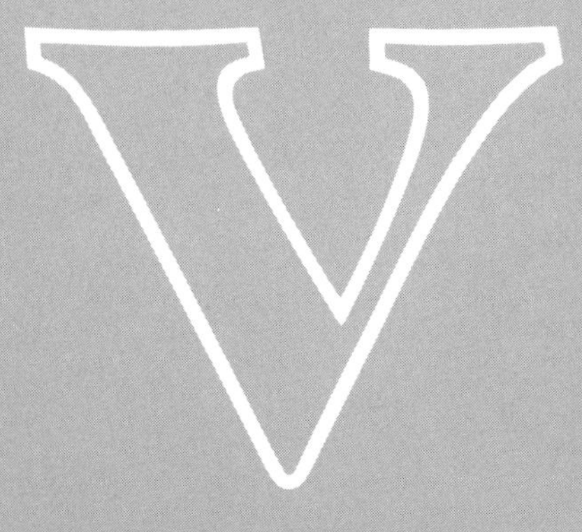

Issues in Multilevel Analysis

Multilevel perspectives on data analysis are essential in the study of complex social realities. However, methodological problems are intensified when there is a simultaneous consideration of group and individual influences on individual behavior. Consequently, many issues emerged around efforts at multilevel analysis. This final part addresses a number of these issues. They include questions about interpreting the contribution of group effects to the explanation of individual behavior, questions involving multilevel theory development, questions about measurement and data requirements, and some technical questions. Our positions on these issues are clarified, and where applicable, we refer specifically to the parts of the book that address the problems of multilevel analysis and the shortcomings of previous work.

Issues in Multilevel Analysis

Chapter 13 focuses on the issues surrounding contextual analysis. Issues involving the explanation of group-level relationships are closely related, because achieving that objective depends on explaining individual behavior through contextual analysis.

Issues are discussed around four general problem areas with descriptive subheadings to identify specific issues. The first problem is one of interpreting the contribution of group effects to the explanation of individual behavior. The issues here deal with problems of stepwise explanation of the total variation, interpreting the importance of variables, and the problem of social selection and migration. The second problem is one of theory development and specification. The issues discussed here are oversimplification with regard to the number of variables considered, oversimplification with regard to the group types and levels considered, and specification of causal connections between group variables and individual behavior. The third problem is one of measurement and data. Issues involving the definition of group variables, the effects of measurement errors, and the data problem are discussed here. Finally, the fourth general problem is one of technical problems and complications.

13.1 *Problems of Interpreting the Contribution of Group Effects to the Explanation of Individual Behavior*

13.1.1. Problems with Stepwise Explained Variance Interpretations of the Importance of Explanatory Variables

One criticism of contextual analysis is aimed specifically at analyses based on the Blau and the Davis approaches for detecting group effects. The criticism is that table and graph representations of relationships exaggerate the importance of the group effect reported in those studies. When the same data are reanalyzed with contemporary analysis of covariance or regression techniques, the reported group effect is described as negligible, trivial, or uninteresting (for example, cf. Schussler, 1969: 230–235; Hauser, 1974:368–371).

These critiques of existing studies must be interpreted carefully. Finding a weak group effect, by whatever method, in one study or a number of studies does not mean that strong group effects do not exist somewhere. Critiques of particular empirical results cannot therefore be taken as a criticism of contextual analysis *per se*. However, there is even more reason to be cautious about these critiques. Linear regression techniques are indisputably more powerful than traditional table analysis. They have the potential to utilize information contained in continuous data and to investigate more complex models of social reality containing many independent variables. However, they do not inherently solve the problem of correlated independent variables.

Two questions require answers after the parameters of a linear model have been estimated. One is how to evaluate the relative importance of the independent variables. The other involves deciding which variables to keep in the model. A common practice is to use the amount of variance explained by each variable, entered last, as a measure of relative importance. For each variable this involves a comparison of R^2 with the variable in the analysis to R^2 without it. This measure of usefulness or importance has been referred to as a measure of independent contribution (cf. Hoffman, 1962).

Another common procedure is to eliminate those variables which do not contribute a statistically significant amount to the explanation when entered last. The pitfalls of applying this procedure without considering the role of each variable in the model and the consequences of removing it from the model are amply brought out in the literature.[1] A variable may be important substantively and indispensible in terms of model estimation, and yet explain very little additional variance when entered last. The consequence of removing it is to bias the estimates for the remaining variables.

Nevertheless, the additional explained variance interpretation of importance predominates in disciplines other than economics, and other interpretations of importance often are overlooked. This is especially true where substantive theories and concepts are less well developed. In the absence of established theories, there are greater pressures to rely on more mechanical approaches to model specification and evaluation. There also may be subtle pressures favoring the simple variance interpretation due to disciplinary predispositions about the primacy of certain types of variables and certain types of explanation. For example, it might simply be assumed that individual-level variables and psychological explanations of behavior are far more important than group variables and sociological explanations of behavior.

The specific meaning of "important" and "unimportant" with reference to the role of variables in linear models is directly pertinent in the criticisms of contextual

[1] See, for example, Ward (1962: 74–76); Blalock (1963); Goldberger (1964: 201); Darlington (1968: 165–170); Gordon (1968); Becker (1972); Farkas (1974: 353–355).

analysis. The meaning of "negligible" or "uninteresting" in these critiques refers to the small contribution of the group variable to the explanation of the dependent variable *after* subtracting the contribution of the individual-level variable. The decision to enter the individual variable first is not defended on theoretical grounds, and may reflect a simple preference for individual-level explanations. The group and interaction variables in these studies do explain substantial amounts of variance when they are considered first. Yet the possibility that the group variable is substantively important or necessary for predictive or policy purposes generally is not considered (cf. Cain and Watts, 1972:231–234; Darlington, 1968:169). Also not generally considered is the possibility that the models of individual behavior implicit in the studies need the group variable for reasons other than to explain variance.

Examples throughout the book illustrate how the group-mean variable is necessary for unbiased estimation of the coefficient for the individual-level variable. Awareness of this source of bias is essential in studies of individual behavior. However, it is also central to the explanation of group-level relationships and the problem of cross-level inference. Specifically, models of individual behavior which neglect the effects of group variables will not be consistent with their aggregated counterparts. Putting it another way, incorrectly specified models of individual behavior are a primary cause of aggregation problems and problems of cross-level inference.

13.1.2. The Problem of Social Selection and Migration

Any theory which purports to explain an individual-level behavior in terms of a group property is open to an alternative hypothesis that the apparent effect is really due to selective migration into groups (cf. Hauser, 1974:373–374). Selection can take place in different ways with different implications for model construction and interpretation. The major question is whether individuals select on the independent or the dependent variable.

An example is in the study of mental illness and socioeconomic class. One theory is that the social disorganization of neighborhoods associated with lower class socioeconomic factors increases the probability of certain types of mental illness. The rival theory is that individuals with greater personal tendencies toward mental illness socially select into lower class neighborhoods. This is an example of selecting on the independent variable. In this case the social disorganization theory is tested with a control for the individual-level variable.

However, what if mentally ill individuals sought out communities to be with other mentally ill people? This would be an example of selecting on the dependent variable. The implicit arrow between the dependent variable and the group variable, in this case, points to the group variable, and a control for the individual variable cannot repudiate the claim of an effect from the group variable.

Concern for the selection problem in contextual analysis is well advised. Awareness is essential to avoid incorrect inferences about the existence of group effects. However, the problem of determining the direction of causal effects is not unique to models of behavior containing group variables. In the study of socioeconomic class and mental illness, one theory is that lower socioeconomic status increases the probability of mental illness. On the other hand, it could be that mental illness leads to lower socioeconomic status. In these situations, arguments generally are based on such things as the temporal ordering of observations and substantive knowledge. Methods for estimating reciprocal causal paths do exist and are becoming more widely used in fields like economics, sociology, and political science (cf. Asher, 1974; Duncan, 1970).

13.2 *Problems of Theory Development and Specification*

13.2.1. Oversimplification with Regard to the Number of Variables Considered

A general criticism of early contextual analyses is that they oversimplify reality. A specific criticism is that the model omits important variables (cf. Hauser, 1974:371–372). At early stages of development, all theories oversimplify reality. That is the purpose of abstraction in science. An implication that oversimplification is a criticism of contextual analysis *per se* is in error. There is nothing in contextual analysis which logically precludes the consideration of other variables. Useful critiques suggest *how* specific theories ought to be expanded or elaborated. Extensions of the basic model of contextual analysis to include more variables are described in Chapter 6.

The ultimate need to elaborate and refine contextual theories is assumed. But this objective ought to be distinguished from specific claims that controls for additional individual-level variables show that the group variable is unnecessary. This is but a variation on the argument based on the stepwise explained variance interpretation of importance where the group variable is entered last. The group variable is pitted against a number of individual-level variables and, as expected, its contribution to explained variance is reduced to the extent that it is correlated with the individual variables. In no way should this *automatically* mean that the group variable ought to be removed from the model. If the group variable is necessary, its omission will cause the estimates of the other variables to be biased. Models which are incorrectly specified in this way present a major obstacle to the explanation of aggregate relationships in terms of individual behavior.

The argument for considering other variables applies no less to group variables. It can be assumed that some of the variation that is not explained by a particular model is due to unspecified group-level variables. To avoid attributing all the additional variance explained by individual-level variables to the properties of individuals, each individual-level variable ought to be accompanied by its corresponding group-mean variable. This is why the extensions of the basic model described in Chapter 6 specify *sets* of variables, each consisting of an individual variable, a group variable, and an interaction variable.

13.2.2. Oversimplification with Regard to the Group Types and Levels Considered

Another specific concern for model oversimplification is aimed at the treatment of only *one* group source of influence. In reality, individuals are members of several groups and are subject to their influence. With few notable exceptions, contextual analyses have not dealt with this reality.[2] This is primarily a comment on the state of theory development. It does not suggest a flaw in the approach. Nothing logically precludes the treatment of more than one group level or group type. Extensions of the basic model in this direction are described in Chapter 7.

The need to elaborate explanations of individual behavior to include multiple-group types and levels is different than the need to treat more group variables. Many group variables may be involved with one group type, such as neighborhoods. Similarly, only one group variable may be considered with many group levels or types. Generally the same group variable will have different values for different group types and levels.

The existence of multiple sources of group effects has raised questions about the "correct" or "most important" one to be considered. Early stages of theory development may simply involve a search for promising sources of group effects. Analysis-of-variance and analysis-of-covariance approaches to multilevel explanations of individual behavior are useful at this stage. (This point is made by Firebaugh, 1978:7.) As in the case of adding variables, the need to elaborate models of individual behavior to include multiple sources of group effects is taken for granted. This ought to be distinguished from specific criticisms of early contextual analyses which show that a purported group effect from one source, such as schools, vanishes when controlling for the same group variable from another source, such as peer groups (cf., for example, Campbell and Alexander, 1965). The implication is that the former group type ought not to be retained in explaining individual behavior.

These criticisms are typically based on a stepwise interpretation of importance where a higher level of aggregation, such as schools, is confronted with a lower level, such as peer groups. The argument is that the former contributes very little after controlling for the latter. Since any particular group variable measured at the two levels is bound to be correlated, the same questions raised above ought to be considered and the same caution ought to be taken when drawing conclusions.

13.2.3. Specification of Causal Connections between Group Variables and Individual Behavior

It is observed in Section 1.5 that the analysis-of-covariance approach to the explanation of individual behavior produces an estimate of *unspecified* group effects. Con-

[2] "Components analysis" is an exception involving the analysis-of-variance approach to the explanation of individual behavior. This involved layered or nested groups (cf. Kish, 1962; Stokes, 1965). The case involving overlapping groups is implicit in a number of studies of school and peer-group influences on school children (cf., for example, Campbell and Alexander, 1965).

sider a study of voter support for a racially biased office seeker and personal racial attitudes, where a significant community effect is detected. What, might be asked, is it about communities that affects a person's vote? Is it the soil, the water, the air? By specifying group variables, contextual analysis seeks to identify what that something is. Imagine that a measure of personal racial attitudes is based on a racism scale. Community averages on this scale become a measure of "racial climate" presumably predicting voting behavior independent of personal racism.

It is still reasonable to ask *how* group properties like "racial climate" affect individuals. This question has several different meanings. One concerns the measurement and validity of concepts like "racial climate." This is discussed in the next section. Another meaning of the question is related to the issue involving multiple-group memberships. What social groupings smaller than communities might account for the community effect on individuals? Possibilities include neighborhoods, organizations, work groups, peer groups, and families. From this angle, the question of how community racial climate affects a particular behavior is a request to elaborate the relationship between racial climate at the community level and that behavior.

Specifying models with more than one type or level of group is implied. Observations at the neighborhood level might be considered along with the community level. Findings might indicate the existence of a direct neighborhood effect, a direct community effect, and a community effect through the neighborhood variable. Findings also might suggest that the community effect was an artifact of neighborhood effects. However, any unexamined assumption that higher level group effects always can be explained away by specifying lower level groups is simply another form of reductionism.

How such things as racial climate affect individuals has yet another meaning. It is reflected in questions as: Is it the structure and flow of information in groups that mediates a group effect? Is it explicitly the media or interpersonal communication networks? Is it interpersonal influence processes or social pressures toward conformity that mediate group effects? How are these ideas operationalized? How do individuals actually receive messages or pressures? Implicit in these questions is a request for social, political, or psychological theories about the causal mechanisms that link individual and group properties.[3] A well deserved criticism of contextual analyses is that they have not satisfactorily dealt with this issue (cf. Cox, 1969; Hauser, 1974: 366–367).

The issue of specifying causal theories which explain how individuals are affected by group properties is intimately related to the issue of specifying group types and levels. It well may be that *different* theories are necessary at different levels and in different types of groups. Perhaps information and diffusion theories are relevant when large geographical units are being considered. Social psychological theories about pressures toward uniformity may be more relevant in small groups. Size is not

[3] See, for example, Alissi (1965); Biddle and Thomas (1966); Cartwright and Zander (1968); Coleman (1958–1959, 1964); Collins and Guetzkow (1964); Cox (1969); Festinger et al. (1950); Guetzkow (1968); Hare et al. (1965); Homans (1964); Hyman and Singer (1968); Lindzey and Aronson (1969); Lott and Lott (1961); Mannheim (1966); McKeachie (1954); Merton (1957); Mills (1967); Thibaut and Kelly (1959); Verba (1961); Walker and Heyns (1962).

the only consideration. Formal groups may suggest a different theory than informal groups. Stable groups may suggest a different theory than unstable groups. Homogeneous groups may suggest something different than heterogeneous groups.

The mismatching of mediating theories and group levels affects outcomes and interpretations of effects. Take the possibility that individuals are induced into voting for a racially biased office seeker primarily through the interpersonal influence of neighbors. However, states are studied rather than neighborhoods. Aggregating neighborhoods to the level of states is likely to reduce the variance of the group variable (proportion racist), thereby attenuating the estimates of group and interaction effects. See Section 9.1 for a discussion of this issue.

Another question related to specifying the links between groups and individuals concerns the psychological mechanisms by which individuals respond to group norms and values. One position is that contextual variables involving such things as racist norms ought to be measured in terms of subjective *perceptions* of individuals about their groups' norms rather than by summarized attitudes within groups (cf. Barton, 1968; Flinn, 1970). That one measure is necessarily better than the other cannot be assumed.

It is possible for individuals to misperceive greatly their groups' norms, although that presumably is what will motivate them. On the other hand, these observations may be considerably less reliable. The ideal is to have both measures. The extent to which they measure the same things functions as a reliability check on the concept of prevailing racial values. The extent to which they measure different things can be exploited to refine a theory of individual behavior in terms of social context and psychological mechanisms. Expectedly, this issue comes up again under the topic of measurement problems.

13.3 *Problems of Measurement and Data*

13.3.1. Issues Involving the Definition of Group Variables

The basic model of contextual analysis specifies a group variable based on the group means or proportions for an individual-level variable.[4] This is taken as a measure of *prevailing* attitudes, values, or behaviors. The use of this type of variable has provoked considerable discussion about its validity and substantive meaning. Complaints about the group-mean variable are often requests for a specification of causal mechanisms as described above. For example, the meaning of the group-mean variable is often tied to the hypothesis of direct structural effects (cf., for example,

[4] Types of group variables are described in Cattell et al. (1953); Lazarsfeld and Menzel (1961); Riley (1964); Selvin and Hagstrom (1963); Eulau (1969); Cartwright (1969). A summary of these usages is contained in Firebaugh (1977).

Campbell and Alexander, 1965). The meaning of this variable also is confused with the issue of its usefulness in the framework of explained variance (cf., for example, Hauser, 1974).

Nevertheless, the question is a serious one. Does the group-mean variable measure some meaningful property of groups, and how well does it measure that property? These questions should not be answered with sweeping generalizations. As with any other type of variable, they will vary from case to case. The group-mean variable may be distant from a meaningful concept, yet measure it well. It also can be very close to a substantively meaningful concept and not measure it very well. An example of the first eventuality is contained in Section 6.1.4 where census data on illiteracy and nativity are analyzed to illustrate an extension of contextual analysis. The group variable "proportions of foreign-born" is a strong predictor of individual-level illiteracy. But the meaning of that variable is not at all obvious. Further inquiry indicated that what was producing the effect was the quality and availability of public education in the states. The proportion of foreign-born was acting as a proxy because individuals of foreign birth, at the time of the 1930 census, tended to be located in geographical areas where public education was more available or of better quality. This was observed in Hanushek et al. (1974).

The group-mean variable can also be close to a substantively meaningful concept and not measure it well. For example, in the case of "racial climate," the averages on the racist scale may be only one indicator of the more complex concept. This should prompt a search for other relevant indicators, such as the number of racist organizations, state positions on other racially relevant issues, measures of employment discrimination, and so forth.

A discussion of group variables other than the group means is presented in Section 3.4. As always, theory should guide the selection and definition of variables. Take, for example, a study of the voting behavior of legislators, party identification, and the partisan composition of legislatures. If one's theory suggests that party competition rather than party dominance in legislative bodies is what affects law-making behavior, then the proportion Republican or Democrat is not the appropriate variable. Instead, the ratio of Republicans to Democrats might be more appropriate. If the study involved the effects of legislative customs or norms, one's theory might prescribe the use of legislators' *perceptions* of these norms rather than (or with) summary measures of attitudes.

13.3.2. The Effect of Measurement Error

It has been pointed out that measurement error differentially affects estimates of individual and group effects (cf. Hauser, 1974:372–373). When the individual and group effects have the same sign, random measurement error in the individual variable will inflate the estimate of the individual effect and deflate the estimate of the group effect. When they have opposite signs, the estimate of the group effect will be inflated. The effects of nonrandom measurement are potentially more serious and unpredictable. The point is to be especially careful when relying on single-item measures of individual-level variables.

13.4
The Data Problem

We have emphasized that there are no logical constraints to the extension of contextual analysis to more complex problems. However, there are usually severe data constraints. A contextual analysis generally requires a representative sample of groups and representative samples of individuals within groups. Unless specifically provided for, this excludes the use of most broadscale sample survey data. When provided by design, sample sizes and costs can expectedly be larger. Undoubtedly this has been a significant factor in the relatively slow development of multilevel methodology.

Often something can be done without complete data. This essentially involves combining incomplete data sources.[5] A number of specific formulas for combining various data sources are presented in Chapter 11. An illustration is presented in Chapter 12 where aggregate election data for constituencies are used in conjunction with sample survey data to study the "puzzle of uniform election swings" in Britain.

13.5
Technical Problems and Complications

A number of technical complications emerge in applying contextual analysis procedures. One specifically involves the case of contextual analysis for categorical data. This case is described in Chapter 5. Because the dependent variable is categorical, there is an issue about the propriety of using dummy variables and least-squares estimation procedures. The problem is that least-squares estimates will be unbiased, but not minimum variance estimates. Whether the implications are serious is debatable. Nevertheless, there are technically more suitable estimation procedures for this special case (cf., for example, Nerlove and Press, 1973).

Another complication for estimation occurs when groups vary substantially in size and homogeneity. The implications for estimation and procedures to compensate for this are described in Appendices B and F.

The question of statistical inference in multilevel analysis has been given very little attention. Given multiple levels of analysis with different units, it is difficult to ascertain appropriate n's and degrees of freedom. As in any data analysis, the problem arises mostly with small numbers of observations. There is no attempt in this book to deal with this question.

Difficulties in performing multilevel analyses also are encountered in data processing. This is a consequence of having observations at more than one level. Variable

[5] The possibility of combining data sources was anticipated by Stokes (1969); Hardor and Pappi (1969); Converse (1969); Linz (1969).

transformations, such as the construction of group-mean variables, are generally required. Data sets must often be merged. Familiar social-science computer analysis packages do not always specifically provide for the necessary operations. This is more of a nuisance than a serious problem. In Appendix E we describe a method for doing multilevel analyses with large numbers of observations directly from tabulated data. Illustrations of specific computer operations for multilevel analysis, using a familiar statistical package (SPSS), are contained in Appendix A.

13.6
Summary

A number of methodological issues involving multilevel analysis are discussed in this chapter. These include questions about interpreting the contribution of group effects to the explanation of individual behavior, questions involving multilevel theory development, questions about measurement and data requirements, and some technical questions. Our positions on these issues are clarified, and where applicable, we refer specifically to the parts of this book that address the inherent problems of multilevel analysis and the shortcomings of previous work.

Appendix

Computer Applications for Multilevel Analysis

The computer runs presented here typify applications in multilevel analysis. For reasons of general availability and widespread use, SPSS (Statistical Package for the Social Sciences) is used to illustrate operations. Since specific control cards for SPSS are system dependent, complete job specifications are not attempted. It is assumed that users have a working knowledge of available SPSS packages.

Illustrations are based on the general-purpose subprograms in SPSS. The objective is to highlight the logic of multilevel computer operations and facilitate adaptations to statistical packages other than SPSS.

SPSS has a special-purpose subprogram called AGGREGATE written expressly for multilevel-analysis tasks. This subprogram is described later. It is recommended when extensive multilevel analysis is anticipated and when there are large numbers of groups.

All subsequent examples are based on the data contained in Table 1.2. These data apply to the hypothetical study of social deviance and religiosity. Each illustration consists of a job setup, computer output, and a discussion of the operations.

1

Illustration 1. Regression Involving Individual-level Variables in a Single Subgroup

This example illustrates the selection and analysis of a subset of cases for a particular group. The VARIABLE LIST card contains a variable which identifies the group to which each case belongs. In this example, that variable is referred to as GROUP and consists of the numbers corresponding to the community numbers in the deviance study. The analysis involves the model

$$y_{ik} = d_{0k} + d_{1k}x_{ik} + f_{ik} \tag{A1}$$

This requires a simple regression of the individual variable Y (OPPOSED) on the individual variable X (RELIG), where k equals 1. The SPSS subprogram REGRESSION performs this computer analysis.

In this instance, regression is used. However, any SPSS operations could be performed on the selected subgroup. For example, the SPSS subprogram CONDESCRIPTIVE might be used to obtain the means and variances of the individual-level variables in the VARIABLE LIST. The systematic use of subsetting operations is a distinguishing feature of multilevel computer analysis.

```
RUN NAME            REGRESSION IN A SINGLE SUBGROUP
VARIABLE LIST       OPPOSED, RELIG, GROUP
INPUT MEDIUM        CARD
INPUT FORMAT        FIXED (F2.0, 1X, F2.0, 1X, F1.0)
SELECT IF           (GROUP EQ 1)
REGRESSION          VARIABLES = OPPOSED, RELIG/
                    REGRESSION = OPPOSED WITH RELIG
READ INPUT DATA
 0  0  1            (data cards continued)  30   3  3
10  0  1                                    35   3  3
15  0  1                                    50   3  3
25  2  1                                    35   4  3
 5  3  1                                    30   6  3
45  3  1                                    60   6  3
15  4  1                                    35   8  3
10  6  1                                    75   9  3
30  6  1                                    65   9  3
45  6  1                                    85   9  3
10  1  2                                    35   4  4
20  1  2                                    35   4  4
30  1  2                                    80   4  4
20  2  2                                    70   6  4
15  4  2                                    40   7  4
45  4  2                                    95   7  4
25  6  2                                    75   8  4
30  7  2                                    85  10  4
50  7  2                                    90  10  4
55  7  2                                    95  10  4
FINISH
```

Computer Output

```
OUTPUT 1

* * * * * * * * * * * *M U L T I P L E   R E G R E S S I O N* * * * * * * * * * * *

VALID CASES             10
```

```
DEPENDENT VARIABLE..       OPPOSED

 MEAN RESPONSE            20.00000         STD. DEV.          15.81139


 MULTIPLE R                .47892          ANALYSIS OF VARIANCE      DF
 R SQUARE                  .22937          REGRESSION                 1.
 ADJUSTED R SQUARE         .13304          RESIDUAL                   8.
 STD DEVIATION           14.72213


 SUM OF SQUARES          MEAN SQUARE            F            SIGNIFICANCE
      516.07143           516.07143          2.38105                 .161
     1733.92857           216.74107

 VARIABLE          B          STD ERROR B          F              BETA
                                             ------------     ----------
                                             SIGNIFICANCE

 RELIG        3.0357143       1.9673272      2.3810505          .4789207
                                                  .161
 (CONSTANT)  10.892857        7.5171466      2.0997983
                                                  .185
```

* *

2

Illustration 2. Regression Involving Individual-level Variables in a Sample Consisting of Several Groups ("the Population")

The analysis in this example is based on the model

$$y_{ik} = d_{0p} + d_{1p} x_{ik} + e_{ik} \tag{A2}$$

This is done with the SPSS subprogram REGRESSION, where the individual variable Y (OPPOSED) is regressed on X (RELIG) for all 40 cases in the data set.

Because subsetting operations are not involved in this run, and because group-level variables are not involved, this run does not by itself constitute a multilevel computer analysis. However, there are occasions when a multilevel analysis requires the computation of statistics which apply to all the groups in the analysis taken together. For example, the regression coefficients obtained may measure an aggregate relationship that is to be explained through a contextual analysis of individual behavior. See Chapters 2, 8, and 9. Population statistics are also used in contextual analyses with incomplete data, as in the case involving group data and sample population data. For a discussion of problems involving incomplete data, see Chapters 10 through 12.

```
RUN NAME           REGRESSION WITH ALL GROUPS INCLUDED
VARIABLE LIST      OPPOSED, RELIG
INPUT MEDIUM       CARD
INPUT FORMAT       FIXED (F2.0, 1X, F2.0)
N OF CASES         40
REGRESSION         VARIABLES = OPPOSED, RELIG/
                   REGRESSION = OPPOSED WITH RELIG
READ INPUT DATA
DATA (see Illustration 1 for data cards)
FINISH
```

Computer Output

```
OUTPUT 2

* * * * * * * * * * * M U L T I P L E   R E G R E S S I O N * * * * * * * * * * *

VALID CASES                40

DEPENDENT VARIABLE..       OPPOSED

 MEAN RESPONSE            42.50000          STD. DEV.            26.84237


 MULTIPLE R                .74852          ANALYSIS OF VARIANCE      DF
 R SQUARE                  .56029          REGRESSION                 1.
 ADJUSTED R SQUARE         .54871          RESIDUAL                  38.
 STD DEVIATION           18.03210


 SUM OF SQUARES          MEAN SQUARE            F              SIGNIFICANCE
    15744.04762          15744.04762         48.41989                  .000
    12355.95238            325.15664

 VARIABLE         B           STD ERROR B          F              BETA
                                              ------------    ----------
                                              SIGNIFICANCE

 RELIG       6.8452381      .98373172       48.419886           .7485228
                                               .000
 (CONSTANT)  8.2738095      5.6852545       2.1179281
                                               .154

* * * * * * * * * * * * * * * * * * * * * * * * * * * * * * * * * * * * * * * * *
```

3

Illustration 3. Regression with Aggregate Variables

The model under consideration is

$$\bar{y}_k = d_{0g} + d_{1g}\bar{x}_k + e_k \tag{A3}$$

This requires a simple regression of the group-mean variables obtained from the individual-level variables Y (OPPOSED) and X (RELIG). Given four communities, there are four values for each variable. This analysis requires two computer runs. The first is similar to the setup in Illustration 1 where the SELECT IF card is used to sort out the cases belonging to a particular group. In this example, that operation is performed once for each community using the variable GROUP to identify the cases belonging to each. The subprogram CONDESCRIPTIVE is used after each subsetting operation to obtain the values of the group means of X (OPPOSED) and Y (RELIG).

The values of the groups means are the inputs for a second run in which the regression analysis is performed. This is done with the subprogram REGRESSION. If the groups in this type of analysis are of unequal size, a weighted regression may be appropriate.

```
RUN NAME            REGRESSION WITH GROUP MEAN VARIABLES
                    (PART 1)
VARIABLE LIST       OPPOSED, RELIG, GROUP
INPUT MEDIUM        CARD
INPUT FORMAT        FIXED (F2.0, 1X, F2.0, 1X, F1.0)
N OF CASES          40
READ INPUT DATA
DATA (see Illustration 1 for data cards)
*SELECT IF          (GROUP EQ 1)
CONDESCRIPTIVE      OPPOSED, RELIG
STATISTICS          1
*SELECT IF          (GROUP EQ 2)
CONDESCRIPTIVE      OPPOSED, RELIG
STATISTICS          1
*SELECT IF          (GROUP EQ 3)
CONDESCRIPTIVE      OPPOSED, RELIG
STATISTICS          1
*SELECT IF          (GROUP EQ 4)
CONDESCRIPTIVE      OPPOSED, RELIG
STATISTICS          1
FINISH
```

Computer Output

```
OUTPUT 3(a)

* * * * * * * * * * * * * * * * * * * * * * * * * * * * * * * * * * * * * * * * *

                  *SELECT IF      (GROUP EQ 1)

 VARIABLE  OPPOSED

 MEAN         20.000
 VALID CASES      10     MISSING CASES    0

 VARIABLE RELIG

 MEAN          3.000
 VALID CASES      10     MISSING CASES    0

- - - - - - - - - - - - - - - - - - - - - - - - - - - - -

                  *SELECT IF      (GROUP EQ 2)

 VARIABLE OPPOSED

 MEAN         30.000
 VALID CASES      10     MISSING CASES    0

 VARIABLE RELIG

 MEAN          4.000
 VALID CASES      10     MISSING CASES    0

- - - - - - - - - - - - - - - - - - - - - - - - - - - - -

                  *SELECT IF      (GROUP EQ 3)

 VARIABLE OPPOSED

 MEAN         50.000
 VALID CASES      10     MISSING CASES    0

 VARIABLE RELIG

 MEAN          6.000
 VALID CASES      10     MISSING CASES    0

- - - - - - - - - - - - - - - - - - - - - - - - - - - - -

                  *SELECT IF      (GROUP EQ 4)

 VARIABLE OPPOSED

 MEAN         70.000
 VALID CASES      10     MISSING CASES    0

 VARIABLE RELIG

 MEAN          7.000
 VALID CASES      10     MISSING CASES    0

* * * * * * * * * * * * * * * * * * * * * * * * * * * * * * * * * * * * * * * *
```

```
RUN NAME          REGRESSION WITH GROUP MEAN VARIABLES
                  (PART 2)
VARIABLE LIST     YBAR, XBAR
INPUT MEDIUM      CARD
INPUT FORMAT      FIXED (F2.0, 1X, F1.0)
N OF CASES        4
REGRESSION        VARIABLES = YBAR, XBAR/
                  REGRESSION = YBAR WITH XBAR
READ INPUT DATA
20  3
30  4
50  6
70  7
FINISH
```

Computer Output

```
OUTPUT 3(b)

* * * * * * * * * * * M U L T I P L E   R E G R E S S I O N * * * * * * * * * * *

VALID CASES              4

DEPENDENT VARIABLE..     YBAR

 MEAN RESPONSE          42.50000        STD. DEV.          22.17356


 MULTIPLE R               .98806         ANALYSIS OF VARIANCE      DF
 R SQUARE                 .97627         REGRESSION                 1.
 ADJUSTED R SQUARE        .96441         RESIDUAL                   2.
 STD DEVIATION           4.18330


 SUM OF SQUARES         MEAN SQUARE            F            SIGNIFICANCE
     1440.00000          1440.00000        82.28571                 .012
       35.00000            17.50000


 VARIABLE           B         STD ERROR B          F               BETA
                                               -----------      ----------
                                               SIGNIFICANCE

 XBAR          12.000000     1.3228757      82.285714            .9880644
                                                 .012
 (CONSTANT) -17.500000       6.9372185      6.3636364
                                                 .128

* * * * * * * * * * * * * * * * * * * * * * * * * * * * * * * * * * * * * * * *
```

This is a multilevel computer application because it requires summarizing of individual-level observations to obtain aggregate variables. The regression coefficients obtained may also measure an aggregate relationship which is to be explained through a contextual analysis of individual behavior. See Chapters 2, 8, and 9. The coefficients obtained from the regression of aggregate (mean) variables are also used in multilevel analyses with incomplete data. See Chapters 10 through 12.

Extensions to more complex problems are straightforward. Other group-mean variables can be constructed from other individual-level variables. Group variables not derived from individual-level observations can be directly introduced in the regression analysis. Interactions between aggregate variables can be analyzed by constructing interaction variables based on the products of the variables. For example, if the type of enforcement structure (LAW) for the four communities were introduced to the analysis, the interaction variable involving LAW and XBAR is constructed by the COMPUTE statement as follows:

```
COMPUTE     INTERACT = LAW * XBAR
```

Extensions to nonlinear models are treated similarly. For example, a squared variable is constructed from

```
COMPUTE     XBARSQR = XBAR ** 2
```

The subsetting operations described in this illustration can also be used to compute the between-group and within-group sum of squares of X. The ratio ETA SQUARE based on these sums of squares is required in the explanation of aggregate relationships in terms of individual-level behavior and in contextual analysis with incomplete data. See Chapters 8 through 12. The within sum of squares can be obtained from Part 1 of the above computer analysis by changing the STATISTICS card after each CONDESCRIPTIVE statement to read

```
STATISTICS     1, 6
```

This gives the means and variances of each variable specified in the CONDESCRIPTIVE card for each group. The sum of squares for particular groups equals the variance of X (RELIG) times the number of cases in the groups minus 1. The within sum of squares is the sum of these sums. The between-group sum of squares can be obtained in Part 2 of the above illustration using a CONDESCRIPTIVE and STATISTICS card. The between sum equals the variance of XBAR times the number of groups minus 1. The total sum of squares equals the between sum plus the within sum. ETA SQUARE equals the between sum divided by the total sum.

When the number of groups in a multilevel analysis is large, the SPSS subprogram AGGREGATE is recommended. See Illustration 11.

4

Illustration 4. Analysis-of-variance Approach to Explaining Individual-level Variables (General Linear model)

The general linear model for this analysis is

$$y_{ik} = a_0 + a_2 z_{2ik} + a_3 z_{3ik} + a_4 z_{4ik} + e_{ik} \tag{A4}$$

The first task is to construct $k - 1 = 3$ dummy variables based on the variable GROUP. This is accomplished by the IF statements. One dummy variable (for the first group) is omitted as required in dummy-variable regression. The three dummy variables are regressed on the individual-level variable Y (OPPOSED). Note that the individual-level variable RELIG is not considered in the regression analysis. It appears in the VARIABLE LIST so that the same data cards and input format as in Illustration 1 can be used.

```
RUN NAME          ANALYSIS OF VARIANCE APPROACH
VARIABLE LIST     OPPOSED, RELIG, GROUP
INPUT MEDIUM      CARDS
INPUT FORMAT      FIXED (F2.0, 1X, F2.0, 1X, F1.0)
N OF CASES        40
IF                (GROUP EQ 2) G2 = 1
IF                (GROUP EQ 3) G3 = 1
IF                (GROUP EQ 4) G4 = 1
REGRESSION        VARIABLES = OPPOSED, G2, G3, G4/
                  REGRESSION = OPPOSED WITH G2, G3, G4
READ INPUT DATA
DATA (see Illustration 1 for data cards)
FINISH
```

Computer Output

```
OUTPUT 4

* * * * * * * * * * * M U L T I P L E   R E G R E S S I O N * * * * * * * * * * * *

VALID CASES              40

DEPENDENT VARIABLE..     OPPOSED
```

```
MEAN RESPONSE          42.50000          STD. DEV.          26.84237

MULTIPLE R              .72451           ANALYSIS OF VARIANCE     DF
R SQUARE                .52491           REGRESSION                3.
ADJUSTED R SQUARE       .48532           RESIDUAL                 36.
STD DEVIATION         19.25703

SUM OF SQUARES          MEAN SQUARE          F             SIGNIFICANCE
   14750.00000          4916.66667        13.25843                 .000
   13350.00000           370.83333

VARIABLE          B          STD ERROR B          F             BETA
                                            SIGNIFICANCE    ----------

G2           10.000000     8.6120071       1.3483146         .1633719
                                                .253
G3           30.000000     8.6120071       12.134831         .4901158
                                                .001
G4           50.000000     8.6120071       33.707865         .8168597
                                                .000
(CONSTANT)   20.000000     6.0896086       10.786517
                                                .002

* * * * * * * * * * * * * * * * * * * * * * * * * * * * * * * * * * * * * * * *
```

5

Illustration 5. Analysis of Covariance Approach to Explaining Individual-level Variables (General Linear Model)

The analysis model in this illustration is

$$y_{ik} = b_0 + b_2 z_{2ik} + b_3 z_{3ik} + b_4 z_{4ik} + b_5 x_{ik} + e_{ik} \tag{A5}$$

The dummy variables for the groups are constructed as in the analysis of variance (Illustration 4). The difference between the analysis-of-variance and the analysis-of-covariance procedures is that the individual-level variable RELIG is included in the REGRESSION cards for the analysis of covariance.

```
RUN NAME          ANALYSIS OF COVARIANCE APPROACH
VARIABLE LIST     OPPOSED, RELIG, GROUP
INPUT MEDIUM      CARD
INPUT FORMAT      FIXED (F2.0, 1X, F2.0, 1X, F1.0)
```

```
N OF CASES          40
IF                  (GROUP EQ 2) G2 = 1
IF                  (GROUP EQ 3) G3 = 1
IF                  (GROUP EQ 4) G4 = 1
REGRESSION          VARIABLES = OPPOSED, RELIG, G2, G3, G4/
                    REGRESSION = OPPOSED WITH RELIG, G2, G3, G4
READ INPUT DATA
DATA (see Illustration 1 for data cards)
FINISH
```

Computer Output

```
OUTPUT 5

* * * * * * * * * * *M U L T I P L E   R E G R E S S I O N* * * * * * * * * * *

VALID CASES               40

DEPENDENT VARIABLE..      OPPOSED

 MEAN RESPONSE          42.50000         STD. DEV.           26.84237


 MULTIPLE R               .84105          ANALYSIS OF VARIANCE      DF
 R SQUARE                 .70737          REGRESSION                 4.
 ADJUSTED R SQUARE        .67393          RESIDUAL                  35.
 STD DEVIATION          15.32774


 SUM OF SQUARES         MEAN SQUARE            F              SIGNIFICANCE
    19877.11864          4969.27966        21.15132                   .000
     8222.88136           234.93947


 VARIABLE           B            STD ERROR B           F               BETA
                                                 SIGNIFICANCE       ----------


 RELIG        4.6610169        .99775058        21.823147          .5096795
                                                     .000
 G2           5.3389831        6.9270051        .59405420          .0872240
                                                     .446
 G3           16.016949        7.4798028        4.5854219          .2616720
                                                     .039
 G4           31.355932        7.9319602        15.627100          .5122680
                                                     .000
 (CONSTANT)   6.0169492        5.6967976        1.1155553
                                                     .298


* * * * * * * * * * * * * * * * * * * * * * * * * * * * * * * * * * * * * * * * *
```

The possibility of interactions involving the individual and group variables can be analyzed by constructing three interaction variables based on the products of each dummy variable and the individual-level variable X. This is done with the COMPUTE statement as follows:

```
COMPUTE    INTERACT2 = G2 * RELIG
COMPUTE    INTERACT3 = G3 * RELIG
COMPUTE    INTERACT4 = G4 * RELIG
```

These variables are included in the REGRESSION statements.

6

Illustration 6. Contextual Analysis with a Single Equation

This computer analysis is based on the basic model of contextual analysis specified in the equation

$$y_{ik} = b_0 + b_1 x_{ik} + b_2 \bar{x}_k + b_3 x_{ik} \bar{x}_k + e_{ik} \tag{A6}$$

As in Illustration 3 involving the construction and analysis of aggregate variables, this analysis requires two computer runs. The first obtains the values of the group means for the variable X (RELIG). This is accomplished by the SELECT IF operations, using the variable GROUP to identify the cases belonging to each group. The difference between the analysis in Illustration 3 and the analysis in this illustration is in the second run. Instead of working with K values of means, the values of the means are assigned to the individuals within the groups. This is accomplished by the IF statements. All individuals in community 1 get the value of the mean of X for that group, and so forth. The result is a new variable XBAR40 consisting of 40 cases.

The interaction variable IACT is the product of the compositional variable XBAR40 and RELIG obtained in the COMPUTE operation. This is followed by the REGRESSION cards where OPPOSED is regressed on RELIG, XBAR40, and IACT.

```
RUN NAME            CONTEXTUAL ANALYSIS WITH A SINGLE EQUATION
                    (PART 1)
VARIABLE LIST       OPPOSED, RELIG, GROUP
INPUT MEDIUM        CARD
INPUT FORMAT        FIXED (F2.0, 1X, F2.0, 1X, F1.0)
N OF CASES          40
READ INPUT DATA
DATA (see Illustration 1 for data cards)
*SELECT IF          (GROUP EQ 1)
CONDESCRIPTIVE      RELIG
STATISTICS          1
```

```
*SELECT IF          (GROUP EQ 2)
CONDESCRIPTIVE      RELIG
STATISTICS          1
*SELECT IF          (GROUP EQ 3)
CONDESCRIPTIVE      RELIG
STATISTICS          1
*SELECT IF          (GROUP EQ 4)
CONDESCRIPTIVE      RELIG
STATISTICS          1
FINISH
```

Computer Output

```
OUTPUT 6(a)

* * * * * * * * * * * * * * * * * * * * * * * * * * * * * * * * * * * * * * * * * *

                    *SELECT IF       (GROUP EQ 1)

 VARIABLE RELIG

 MEAN           3.000
 VALID CASES       10      MISSING CASES     0

- - - - - - - - - - - - - - - - - - - - - - - - - - - - -

                    *SELECT IF       (GROUP EQ 2)

 VARIABLE RELIG

 MEAN           4.000
 VALID CASES       10      MISSING.CASES     0

- - - - - - - - - - - - - - - - - - - - - - - - - - - - -

                    *SELECT IF       (GROUP EQ 3)

 VARIABLE RELIG

 MEAN           6.000
 VALID CASES       10      MISSING CASES     0

- - - - - - - - - - - - - - - - - - - - - - - - - - - - -

                    *SELECT IF       (GROUP EQ 4)

 VARIABLE RELIG

 MEAN           7.000
 VALID CASES       10      MISSING CASES     0

* * * * * * * * * * * * * * * * * * * * * * * * * * * * * * * * * * * * * * * * * *
```

```
RUN NAME            CONTEXTUAL ANALYSIS WITH A SINGLE EQUATION
                    (PART 2)
```

```
VARIABLE LIST      OPPOSED, RELIG, GROUP
INPUT MEDIUM       CARD
INPUT FORMAT       FIXED (F2.0, 1X, F2.0, 1X, F1.0)
N OF CASES         40
IF                 (GROUP EQ 1) XBAR40 = 3
IF                 (GROUP EQ 2) XBAR40 = 4
IF                 (GROUP EQ 3) XBAR40 = 6
IF                 (GROUP EQ 4) XBAR40 = 7
COMPUTE            IACT = XBAR40 * RELIG
REGRESSION         VARIABLES = OPPOSED, RELIG, XBAR40, IACT/
                   REGRESSION = OPPOSED WITH RELIG, XBAR40, IACT
READ INPUT DATA
DATA (see Illustration 1 for data cards)
FINISH
```

Computer Output

```
OUTPUT 6(b)

* * * * * * * * * * * M U L T I P L E   R E G R E S S I O N * * * * * * * * * * *

VALID CASES               40

DEPENDENT VARIABLE..      OPPOSED

 MEAN RESPONSE          42.50000          STD. DEV.          26.84237


 MULTIPLE R               .84652          ANALYSIS OF VARIANCE       DF
 R SQUARE                 .71659          REGRESSION                  3.
 ADJUSTED R SQUARE        .69297          RESIDUAL                   36.
 STD DEVIATION          14.87334


 SUM OF SQUARES          MEAN SQUARE            F             SIGNIFICANCE
    20136.21985           6712.07328        30.34170                  .000
     7963.78015            221.21612


 VARIABLE          B          STD ERROR B          F                BETA
                                             ------------       ----------
                                             SIGNIFICANCE

 RELIG       -.13505556     3.0481862       .19630978E-02       -.0147682
                                                 .965
 XBAR40       2.5429105     3.3916958       .56211914            .1516973
                                                 .458
 IACT         .95921450     .57806861       2.7534215            .7364088
                                                 .106
 (CONSTANT)  4.0823263      15.165903       .72456756E-01
                                                 .789

* * * * * * * * * * * * * * * * * * * * * * * * * * * * * * * * * * * * * * * *
```

Additional sets of variables (individual, group, and interaction) are analyzed by the same procedures. Quadratic terms and other variable transformations are made using the COMPUTE statement. Procedures involving more than one group type or level are essentially the same. Instead of one grouping variable, such as GROUP in the simple case, there will be two or more. The case of nested groups requires a series of steps to obtain means at various levels of aggregation and to construct new compositional variables. Procedures for the separate-equation approach to contextual analysis are described in Illustration 7. The centered model is considered in Illustration 8.

After constructing appropriate dummy variables with IF statements, procedures for the case involving dichotomous variables are the same as above. However, when the number of observations in contingency tables is very large, as in census data, the construction of dummy variables may be quite impractical. In this case, correlation coefficients, means, and standard deviations directly calculated from the tables can be used as inputs into regression programs. SPSS has this capability. For details on this procedure, see Appendix E.

While the basic operations described above can be generalized to more complex analyses, they can be tedious. When extensive analysis is anticipated, the SPSS subprogram AGGREGATE in Illustration 11 is recommended.

7

Illustration 7. Contextual Analysis with Separate Equations

The separate-equation approach to contextual analysis involves a two-step procedure. The first requires a regression of the individual-level variable Y (OPPOSED) on the individual-level variable X (RELIG) *within* each of the designated groups. This is accomplished by preceding each regression analysis by an appropriate SELECT IF card. This step outputs the values of the within-group intercepts and the slopes. The SPSS subprogram REGRESSION also outputs the means and standard deviations of the variables included in the VARIABLE LIST. Otherwise an operation comparable to CONDESCRIPTIVE in SPSS is necessary at this point to obtain the values of the group means of X (RELIG) which are used in the second part of the analysis.

The second step requires that the group intercepts, slopes, and means of X be entered as data. These variables are designated in the VARIABLE LIST card as ICEPTS, SLPS, and XBAR. Since there are four communities in the illustration, there are four values for each of these variables.

The separate-equation model specifies the two equations

$$d_{0k} = b_0 + b_2\bar{x}_k + u_k \tag{A7}$$

$$d_{1k} = b_1 + b_3\bar{x}_k + v_k \tag{A8}$$

The first of these equations requires the regression of ICEPTS on XBAR. The second involves the regression of SLPS on XBAR. If the groups vary substantially in size, these regressions should be weighted. For a discussion of this procedure see Appendix B. The coefficients obtained from the separate-equation procedure described here will not necessarily be the same as those obtained from the single-equation procedure described previously. For a discussion of this point, see Appendix C.

```
RUN NAME         CONTEXTUAL ANALYSIS WITH SEPARATE EQUATIONS
                 (PART 1)
VARIABLE LIST    OPPOSED, RELIG, GROUP
INPUT MEDIUM     CARD
INPUT FORMAT     FIXED (F2.0, 1X, F2.0, 1X, F1.0)
N OF CASES       40
READ INPUT DATA
DATA (see Illustration 1 for data cards)
*SELECT IF       (GROUP EQ 1)
REGRESSION       VARIABLES = OPPOSED, RELIG/
                 REGRESSION = OPPOSED WITH RELIG
*SELECT IF       (GROUP EQ 2)
REGRESSION       VARIABLES = OPPOSED, RELIG/
                 REGRESSION = OPPOSED WITH RELIG
*SELECT IF       (GROUP EQ 3)
REGRESSION       VARIABLES = OPPOSED, RELIG/
                 REGRESSION = OPPOSED WITH RELIG
*SELECT IF       (GROUP EQ 4)
REGRESSION       VARIABLES = OPPOSED, RELIG/
                 REGRESSION = OPPOSED WITH RELIG
FINISH
```

Computer Output

```
OUTPUT 7(a)

* * * * * * * * * * * M U L T I P L E   R E G R E S S I O N * * * * * * * * * * *

SELECT IF        (GROUP EQ 1)

VALID CASES            10

DEPENDENT VARIABLE..   OPPOSED

 MEAN RESPONSE       20.00000        STD. DEV.         15.81139

 MULTIPLE R            .47892         ANALYSIS OF VARIANCE     DF
 R SQUARE              .22937         REGRESSION                1.
 ADJUSTED R SQUARE     .13304         RESIDUAL                  8.
 STD DEVIATION       14.72213
```

```
SUM OF SQUARES        MEAN SQUARE          F            SIGNIFICANCE
    516.07143          516.07143          2.38105               .161
   1733.92857          216.74107

VARIABLE          B          STD ERROR B        F            BETA
                                           SIGNIFICANCE   ----------

RELIG        3.0357143     1.9673272     2.3810505       .4789207
                                              .161
(CONSTANT)  10.982857      7.5171466     2.0997983
                                              .185

* * * * * * * * * * *M U L T I P L E   R E G R E S S I O N* * * * * * * * * * *

*SELECT IF      (GROUP EQ 2)

VALID CASES              10

DEPENDENT VARIABLE..     OPPOSED

MEAN RESPONSE          30.00000        STD. DEV.         15.27525

MULTIPLE R               .65127        ANALYSIS OF VARIANCE     DF
R SQUARE                 .42416        REGRESSION                1.
ADJUSTED R SQUARE        .35217        RESIDUAL                  8.
STD DEVIATION          12.29468

SUM OF SQUARES        MEAN SQUARE          F            SIGNIFICANCE
    890.72581          890.72581          5.89263               .041
   1209.27419          151.15927

VARIABLE          B          STD ERROR B        F            BETA
                                           SIGNIFICANCE   ----------

RELIG        3.7903226     1.5614265     5.8926309       .6512719
                                              .041
(CONSTANT)  14.838710      7.3569540     4.0681428
                                              .078

* * * * * * * * * * *M U L T I P L E   R E G R E S S I O N* * * * * * * * * * *

SELECT IF      (GROUP EQ 3)

VALID CASES              10

DEPENDENT VARIABLE..     OPPOSED

MEAN RESPONSE          50.00000        STD. DEV.         20.13841

MULTIPLE R               .69370        ANALYSIS OF VARIANCE     DF
R SQUARE                 .48122        REGRESSION                1.
ADJUSTED R SQUARE        .41637        RESIDUAL                  8.
STD DEVIATION          15.38485
```

```
 SUM OF SQUARES           MEAN SQUARE           F              SIGNIFICANCE
     1756.45161            1756.45161        7.42078                   .026
     1893.54839             236.69355


 VARIABLE          B           STD ERROR B          F                BETA
                                                SIGNIFICANCE      ----------

 RELIG       5.3225806      1.9538777       7.4207836            .6936999
                                                 .026
 (CONSTANT)  18.064516      12.692688       2.0255617
                                                 .192

* * * * * * * * * * * M U L T I P L E   R E G R E S S I O N * * * * * * * * * * *

SELECT IF        (GROUP EQ 4)

VALID CASES                  10

DEPENDENT VARIABLE..         OPPOSED

 MEAN RESPONSE            70.00000          STD. DEV.           24.38123


 MULTIPLE R                .66684          ANALYSIS OF VARIANCE      DF
 R SQUARE                  .44468          REGRESSION                 1.
 ADJUSTED R SQUARE         .37526          RESIDUAL                   8.
 STD DEVIATION           19.27103


 SUM OF SQUARES           MEAN SQUARE           F              SIGNIFICANCE
     2379.01786            2379.01786        6.40601                   .035
     2970.98214             371.37277

 VARIABLE          B           STD ERROR B          F                BETA
                                                SIGNIFICANCE      ----------

 RELIG       6.5178571      2.5752003       6.4060105            .6668405
                                                 .035
 (CONSTANT)  24.375000      19.028622       1.6408715
                                                 .236

* * * * * * * * * * * * * * * * * * * * * * * * * * * * * * * * * * * * * * * * *
```

RUN NAME	CONTEXTUAL ANALYSIS WITH SEPARATE EQUATIONS (PART 2)
VARIABLE LIST	ICEPTS, SLPS, XBAR
INPUT MEDIUM	CARD
INPUT FORMAT	FIXED (F4.1, 1X, F3.1, 1X, F1.0)
N OF CASES	4

```
REGRESSION        VARIABLES = ICEPTS, SLPS, XBAR/
                  REGRESSION = ICEPTS WITH XBAR/
                  REGRESSION = SLPS WITH XBAR
READ INPUT DATA
10.9  3.0  3
14.8  3.8  4
18.1  5.3  6
24.4  6.5  7
FINISH
```

Computer Output

```
OUTPUT 7(b)

* * * * * * * * * * * M U L T I P L E   R E G R E S S I O N * * * * * * * * * * *

VALID CASES              4

DEPENDENT VARIABLE..     ICEPTS

 MEAN RESPONSE          17.05000         STD. DEV.          5.71577


 MULTIPLE R              .96785           ANALYSIS OF VARIANCE      DF
 R SQUARE                .93673           REGRESSION                 1.
 ADJUSTED R SQUARE       .90510           RESIDUAL                   2.
 STD DEVIATION          1.76082


 SUM OF SQUARES          MEAN SQUARE              F             SIGNIFICANCE
       91.80900            91.80900          29.61103                   .032
        6.20100             3.10050

 VARIABLE            B          STD ERROR B         F                BETA
                                              ------------       ----------
                                              SIGNIFICANCE

 XBAR         3.0300000      .55682134       29.611030            .9678486
                                                  .032
 (CONSTANT)   1.9000000      2.9199957       .42339212
                                                  .582


* * * * * * * * * * * M U L T I P L E   R E G R E S S I O N * * * * * * * * * * *

VALID CASES              4

DEPENDENT VARIABLE..     SLPS

 MEAN RESPONSE           4.65000         STD. DEV.          1.55885
```

```
MULTIPLE R              .99553         ANALYSIS OF VARIANCE      DF
R SQUARE                .99108         REGRESSION                 1.
ADJUSTED R SQUARE       .98663         RESIDUAL                   2.
STD DEVIATION           .18028


SUM OF SQUARES          MEAN SQUARE            F              SIGNIFICANCE
       7.22500              7.22500        222.30769                  .004
        .06500               .03250


VARIABLE          B            STD ERROR B          F             BETA
                                              SIGNIFICANCE     ----------

XBAR          .85000000      .57008771E-01  222.30769          .9955319
                                                 .004
(CONSTANT)    .40000000      .29895652      1.7902098
                                                 .313
```

* *

With dichotomous individual-level variables, the values of the intercepts, slopes, and group means used in the regression analysis can be obtained directly from the tables. For a discussion see Chapter 5. Extensions of the basic model, as to the case involving nonlinear context effects, involve variations of the procedures described above. For example, a squared variable based on the group means XBAR is constructed by the COMPUTE statement:

COMPUTE XBARSQ = XBAR ** 2

Procedures for the centered model are described in Illustration 9.

8

Illustration 8. Procedures for the Centered Model of Contextual Analysis with a Single Equation

The centered model for contextual analysis described in Chapter 4 is

$$y'_{ik} = a_0 + a_1(x_{ik} - \bar{x}_k) + a_2(\bar{x}_k - \bar{x}) + a_3(x_{ik} - \bar{x}_k)(\bar{x}_k - \bar{x}) + e_{ik} \qquad \text{(A9)}$$

where the adjusted values (y'_{ik}) are obtained from

$$y'_{ik} = y_{ik} - d_{1k}\bar{x}_k \qquad \text{(A10)}$$

This requires two computer runs. The first obtains the within-group slopes d_{1k} and the group means $\bar{x}_k$. Essentially the same logic is involved as in the first part of

Illustration 7 for the separate-equation procedure. The grand mean $\bar{x}$ is also necessary. This requires that all (40) cases in the data be considered as in Illustration 2. It is obtained by a CONDESCRIPTIVE card.

The run shown here assigns the values of the slopes d_{1k} and the means $\bar{x}_k$ to each of the 40 cases. This is accomplished by the IF statements. The new variables are designated SLPS40 and XBAR40, respectively. The new dependent variable is constructed with the first COMPUTE statement, using the variables OPPOSED, SLPS40, and XBAR40. The new explanatory variables are constructed from the remaining COMPUTE statements. The new variables are designated NEWY, NEWX, NEWXBAR, and NEWIACT. This is followed by the REGRESSION cards specifying the regression of the new dependent variable on the three new explanatory variables.

```
RUN NAME          THE CENTERED MODEL PROCEDURES
                  (SINGLE EQUATION)
VARIABLE LIST     OPPOSED, RELIG, GROUP
INPUT MEDIUM      CARD
INPUT FORMAT      FIXED (F2.0, 1X, F2.0, 1X, F1.0)
N OF CASES        40
IF                (GROUP EQ 1) SLPS40 = 3.0
IF                (GROUP EQ 2) SLPS40 = 3.8
IF                (GROUP EQ 3) SLPS40 = 5.3
IF                (GROUP EQ 4) SLPS40 = 6.5
IF                (GROUP EQ 1) XBAR40 = 3
IF                (GROUP EQ 2) XBAR40 = 4
IF                (GROUP EQ 3) XBAR40 = 6
IF                (GROUP EQ 4) XBAR40 = 7
COMPUTE           NEWY = OPPOSED – (SLPS40 * XBAR40)
COMPUTE           NEWX = RELIG – XBAR40
COMPUTE           NEWXBAR = XBAR40 – 5.0
COMPUTE           NEWIACT = NEWX * NEWXBAR
REGRESSION        VARIABLES = NEWY, NEWX, NEWXBAR, NEWIACT/
                  REGRESSION = NEWY WITH NEWX, NEWXBAR, NEWIACT
READ INPUT DATA
DATA (see Illustration 1 for data cards)
FINISH
```

Computer Output

```
OUTPUT 8

* * * * * * * * * * * * M U L T I P L E   R E G R E S S I O N * * * * * * * * * * * *

VALID CASES              40

DEPENDENT VARIABLE..     NEWY

 MEAN RESPONSE         17.12500        STD. DEV.          19.17174
```

```
MULTIPLE R              .67144       ANALYSIS OF VARIANCE      DF
R SQUARE                .45083       REGRESSION                 3.
ADJUSTED R SQUARE       .40507       RESIDUAL                  36.
STD DEVIATION         14.78754


SUM OF SQUARES          MEAN SQUARE          F              SIGNIFICANCE
    6462.51116           2154.17039        9.85118                  .000
    7872.16384            218.67122


VARIABLE           B           STD ERROR B         F                 BETA
                                              SIGNIFICANCE        ----------

NEWX          4.6610169      .96258660      23.446701           .5980572
                                                 .000
NEWXBAR       3.0400000      1.4787536      4.2262535           .2539099
                                                 .047
NEWIACT       .84790210      .61829794      1.8805974           .1693752
                                                 .179
(CONSTANT)    17.125000      2.3381147      53.645034
                                                 .000

* * * * * * * * * * * * * * * * * * * * * * * * * * * * * * * * * * * * * * * * *
```

9

Illustration 9. Procedures for the Centered Model of Contextual Analysis with Separate Equations

The centered model with separate equations is specified in two equations.

$$d_{0k} = a_0 + a_2(\bar{x}_k - \bar{x}) + u_k \tag{A11}$$

$$d_{1k} = a_1 + a_3(\bar{x}_k - \bar{x}) + v_k \tag{A12}$$

This computer analysis requires a two-step procedure. The first step is identical to the first part of Illustration 7 for the uncentered analysis with separate equations. Additionally, the grand mean $\bar{x}$ for all 40 cases must be computed as in Illustration 2.

The second step requires the construction of the adjusted group-mean variable NEWXBAR. This is accomplished by the COMPUTE statement. The group intercepts ICEPTS are then regressed on the adjusted means NEWXBAR, and the group slopes SLPS are regressed on the adjusted means NEWXBAR.

```
RUN NAME          CENTERED MODEL PROCEDURES WITH SEPARATE
                  EQUATIONS
```

```
VARIABLE LIST     ICEPTS, SLPS, XBAR
INPUT MEDIUM      CARD
INPUT FORMAT      FIXED (F4.1, 1X, F3.1, 1X, F1.0)
N OF CASES        4
COMPUTE           NEWXBAR = XBAR − 5.0
REGRESSION        VARIABLES = ICEPTS, SLPS, NEWXBAR/
                  REGRESSION = ICEPTS WITH NEWXBAR/
                  REGRESSION = SLPS WITH NEWXBAR
READ INPUT DATA
10.9  3.0  3
14.8  3.8  4
18.1  5.3  6
24.4  6.5  7
FINISH
```

Computer Output

```
OUTPUT 9

* * * * * * * * * * * * M U L T I P L E   R E G R E S S I O N * * * * * * * * * * * *

VALID CASES                 4

DEPENDENT VARIABLE..        ICEPTS

 MEAN RESPONSE          17.05000          STD. DEV.           5.71577


 MULTIPLE R              .96785           ANALYSIS OF VARIANCE      DF
 R SQUARE                .93673           REGRESSION                 1.
 ADJUSTED R SQUARE       .90510           RESIDUAL                   2.
 STD DEVIATION          1.76082


 SUM OF SQUARES          MEAN SQUARE           F              SIGNIFICANCE
       91.80900             91.80900       29.61103                   .032
        6.20100              3.10050

 VARIABLE          B            STD ERROR B          F              BETA
                                              SIGNIFICANCE       ----------

 NEWXBAR      3.0300000       .55682134       29.611030          .9678486
                                                   .032
 (CONSTANT)  17.050000        .88041184       375.03951
                                                   .003

* * * * * * * * * * * * M U L T I P L E   R E G R E S S I O N * * * * * * * * * * * *

VALID CASES                 4

DEPENDENT VARIABLE..        SLPS
```

```
MEAN RESPONSE            4.65000          STD. DEV.           1.55885

MULTIPLE R               .99553           ANALYSIS OF VARIANCE      DF
R SQUARE                 .99108           REGRESSION                 1.
ADJUSTED R SQUARE        .98663           RESIDUAL                   2.
STD DEVIATION            .18028

SUM OF SQUARES           MEAN SQUARE           F            SIGNIFICANCE
       7.22500               7.22500       222.30769                .004
        .06500                .03250

VARIABLE            B           STD ERROR B            F             BETA
                                                  SIGNIFICANCE    ----------

NEWXBAR       .85000000      .57008771E-01  222.30769            .9955319
                                                  .004
(CONSTANT)   4.6500000       .90138782E-01  2661.2308
                                                  .000
```

* *

10

Illustration 10. Explaining Aggregate Relationships and Contextual Analysis with Incomplete Data

Chapters 8 through 12 describe a number of multilevel analysis procedures involving group data, group specific data, and population data. Computer operations for these analyses involve various combinations of the procedures previously illustrated. To illustrate, consider the model of contextual structure represented in Eq. (8.29). This equation is

$$d_{1p} = b_1(1 - \eta^2) + b_3\bar{x}(1 - \eta^2) + d_{1g}\eta^2 \tag{A13}$$

One use of this equation is to obtain measures of individual, group, and interaction effects when only group data and sample population data are available. Thus the value of the population slope d_{1p} is obtained from a regression with all the observations in the data. This essentially involves the procedure described in Illustration 2. The value of the group slope d_{1g} is obtained in the regression of the group means as described in Illustration 3. The value of the interaction effect is obtained in a multiple regression containing the squared term $\bar{x}_k^2$. This is another application of the procedures for aggregate variables described in Illustration 3. The grand mean $\bar{x}$ is obtained from the population data as in Illustration 2. The value of η^2 is obtained from the group data and the population data by computing the between and total sums of squares. An estimate of the individual effect is obtained by solving for b_1 in Eq. (A13).

An alternative computer approach is provided by the SPSS subprogram AGGREGATE described in Illustration 11. That alternative is especially useful when an analysis of aggregate and individual-level data files is required.

11

Illustration 11. Use of the SPSS Subprogram AGGREGATE for Multilevel Analyses

This subprogram was written in SPSS for multilevel analysis. The computer illustrations presented above depend only on the more familiar general-purpose operations in SPSS. The objective is to complement the discussion of multilevel concepts and statistical procedures presented in this book. However, the SPSS subprogram AGGREGATE is recommended for extended multilevel analyses, especially when they involve complex analysis models or large numbers of groups.

Two types of files are created by the AGGREGATE program. The first constructs aggregate variables from individual-level data and grouping variables. The first job setup below shows the construction of two aggregate variables based on the means of OPPOSED and RELIG for each community. In SPSS this type of file is referred as a *true* file. We refer to this type of file as an *aggregate variable file*. This type is distinguished from a *compositional file*, to be described subsequently. An aggregate-variable file is produced when there is no OPTION card as in the second setup.

The AGGREGATE card does two things. The GROUPVAR part specifies the grouping variable (or variables) to be used in the construction of aggregate variables. In this example it is the variable GROUP identifying the community to which each individual in the study belongs. The VARIABLE LIST of the AGGREGATE card specifies the individual-level variables from which aggregate variables are constructed. In this example there are two such variables, OPPOSED and RELIG. The AGGSTATS card specifies the summarizing operation to be performed within the specified groups. In the example this involves the computation of the means. The number of observations in the individual-level data is 40. Since there are four groups in this analysis, the number of cases for the aggregate variables is four. These values can be written into a new aggregate file or added to an existing aggregate file containing the same groups.

The aggregate-variable file is used for analyses involving aggregate-level variables, as in a regression of group means. Note that equivalent results are obtained using general-purpose operations as in Illustration 3. The subprogram AGGREGATE is recommended when extensive group-level analyses are anticipated or when the number of groups involved is large.

```
RUN NAME          CREATING AN AGGREGATE VARIABLE DATA FILE
VARIABLE LIST     OPPOSED, RELIG, GROUP
INPUT MEDIUM      CARD
N OF CASES        40
INPUT FORMAT      (F2.0, 1X, F2.0, 1X, F1.0)
AGGREGATE         GROUPVARS = GROUP/VARIABLES = OPPOSED, RELIG/
                  AGGSTATS = MEAN
STATISTICS        1, 2
READ INPUT DATA
DATA (see Illustration 1 for data cards)
FINISH
```

The second type of file generated by the AGGREGATE program in SPSS is referred to as a *compositional file*. The setup below illustrates the first step in creating a compositional file. This requires the OPTION card designating option number 4.

In compositional files the values of aggregate variables such as group means are assigned to individual cases. In this illustration the value of the mean of X (RELIG) obtained for the first group is assigned to the 10 individuals in that group, and so forth until all 40 individuals in the study have a score on this new variable. The same procedure applies to the variable OPPOSED since it is in the VARIABLE LIST. The values of the new group-level variables can be written into a new file or added to an original individual-level data file. The resulting file can then be used to perform contextual analyses and other multilevel operations.

The use of the subprogram AGGREGATE for compositional files produces results equivalent to those described in Illustration 6, where general-purpose operations are used. However, it can greatly reduce tedious operations when extensive multilevel analyses are anticipated.

```
RUN NAME          CREATING A COMPOSITIONAL FILE
VARIABLE LIST     OPPOSED, RELIG, GROUP
INPUT MEDIUM      CARD
N OF CASES        40
INPUT FORMAT      FIXED (F2.0, 1X, F2.0, 1X, F1.0)
AGGREGATE         GROUPVARS = GROUP/VARIABLES = OPPOSED, RELIG/
                  AGGSTATS = MEAN
STATISTICS        1, 2
OPTION            4
READ INPUT DATA
DATA (see Illustration 1 for data cards)
FINISH
```

Computer Output

```
OUTPUT 10

* * * * * * * * * * * * * * * * A G G R E G A T E * * * * * * * * * * * * * * * *

         GROUPING VARIABLES...    GROUP

- - - - - - - - - - - - - - - - - - - - - - - - - - - - - - - - - - - - - - - - -

 GROUP-ID          1   TOTAL N        10   GROUP-VALUES

 VARIABLE      VALID N                SUM        MEAN

 OPPOSED            10                          20.00
 RELIG              10                           3.00



- - - - - - - - - - - - - - - - - - - - - - - - - - - - - - - - - - - - - - - - -

 GROUP-ID          2   TOTAL N       10    GROUP-VALUES

 VARIABLE      VALID N               SUM         MEAN

 OPPOSED            10                          30.00
 RELIG              10                           4.00



- - - - - - - - - - - - - - - - - - - - - - - - - - - - - - - - - - - - - - - - -

 GROUP-ID          3   TOTAL N       10    GROUP-VALUES

 VARIABLE      VALID N               SUM         MEAN

 OPPOSED            10                          50.00
 RELIG              10                           6.00



- - - - - - - - - - - - - - - - - - - - - - - - - - - - - - - - - - - - - - - - -

 GROUP-ID          4   TOTAL N       10    GROUP-VALUES

 VARIABLE      VALID N               SUM         MEAN

 OPPOSED            10                          70.00
 RELIG              10                           7.00



- - - - - - - - - - - - - - - - - - - - - - - - - - - - - - - - - - - - - - - - -
```

```
THE CONTENTS OF EACH CASE ON THE AGGREGATED OUTPUT FILE ..

SEQUENTIAL      CREATED VARIABLE
POSITION

        1       AGGREGATION GROUP NUMBER
        2       TOTAL NUMBER OF CASES IN AGGREGATION GROUP
        3       OPPOSED  - MEAN
        4       RELIG    - MEAN

A TRUE AGGREGATED OUTPUT FILE WAS WRITTEN ON FILE BCDOUT   IT CONTAINS
4 AGGREGATED CASES

* * * * * * * * * * * * * * * * * * * * * * * * * * * * * * * * * * * * * * * * *


* * * * * * * * * * * * * * * A G G R E G A T E * * * * * * * * * * * * * * * *

          GROUPING VARIABLES...    GROUP

- - - - - - - - - - - - - - - - - - - - - - - - - - - - - - - - - - - - - - - - -

 GROUP-ID         1   TOTAL N       10   GROUP-VALUES

 VARIABLE      VALID N              SUM        MEAN

 OPPOSED          10                          20.00
 RELIG            10                           3.00



- - - - - - - - - - - - - - - - - - - - - - - - - - - - - - - - - - - - - - - - -

 GROUP-ID         2   TOTAL N       10   GROUP-VALUES

 VARIABLE      VALID N              SUM        MEAN

 OPPOSED          10                          30.00
 RELIG            10                           4.00



- - - - - - - - - - - - - - - - - - - - - - - - - - - - - - - - - - - - - - - - -

 GROUP-ID         3   TOTAL N       10   GROUP-VALUES

 VARIABLE      VALID N              SUM        MEAN

 OPPOSED          10                          50.00
 RELIG            10                           6.00



- - - - - - - - - - - - - - - - - - - - - - - - - - - - - - - - - - - - - - - - -
```

```
GROUP-ID          4    TOTAL N         10    GROUP-VALUES

VARIABLE       VALID N                  SUM        MEAN

OPPOSED            10                              70.00
RELIG              10                               7.00



- - - - - - - - - - - - - - - - - - - - - - - - - - - - - - - - - - - - - - - -

THE CONTENTS OF EACH CASE ON THE AGGREGATED OUTPUT FILE ..

SEQUENTIAL        CREATED VARIABLE
POSITION

         1        AGGREGATION GROUP NUMBER
         2        TOTAL NUMBER OF CASES IN AGGREGATION GROUP
         3        OPPOSED  - MEAN
         4        RELIG    - MEAN

A COMPOSITIONAL OUTPUT FILE WAS WRITTEN ON FILE BCDOUT   IT CONTAINS
40 UNWEIGHTED CASES

* * * * * * * * * * * * * * * * * * * * * * * * * * * * * * * * * * * * * * * * * *
```

Appendix

Weighted Regression for the Separate Equations

The separate model equations are introduced in Eqs. (3.6) and (3.7) and for the centered approach in Eqs. (4.12) and(4.13). Eqs. (3.6) and (3.7) are

$$d_{0k} = b_0 + b_2\bar{x}_k + u_k \tag{B1}$$

$$d_{1k} = b_1 + b_3\bar{x}_k + v_k \tag{B2}$$

One standard requirement in regression analysis is that the variance of the residuals be constant across the range of values of the explanatory variable. This requirement is not satisfied here because the dependent variables are themselves regression coefficients.

Values of the two dependent variables are found by regressing Y on X within the groups:

$$y_{ik} = d_{0k} + d_{1k}x_{ik} + f_{ik}, \qquad i = 1, \ldots, n_k; k = 1, \ldots, K \tag{B3}$$

Assuming simple random sampling and the same constant variance σ^2 for the residuals in each group, then the variances for d_{0k} and d_{1k} become

$$V(d_{0k}) = \frac{\sigma^2}{n_k} + \frac{\bar{x}_k^2\sigma^2}{\sum(x_{ik} - \bar{x}_k)^2} = w_{0k}\sigma^2 \tag{B4}$$

$$V(d_{1k}) = \frac{\sigma^2}{\sum(x_{ik} - \bar{x}_k)^2} = w_{1k}\sigma^2 \tag{B5}$$

Using $1/w_{0k}$ and $1/w_{1k}$ as weights in a weighted regression, it is possible to get more efficient estimators for the b's. For a further discussion see Hanushek (1974).

Instead of using these weights there are reasons for using the number of individuals in each group as a weight for that group. Without weights the sum of the u residuals equals zero, that is,

$$\sum u_k = 0 \tag{B6}$$

Because the residuals are always uncorrelated with the explanatory variables we also have, for the v's,

$$\sum \bar{x}_k v_k = 0 \tag{B7}$$

By using n_k as weight for the kth group the b's are found from the expressions

$$b_2 = \frac{\sum n_k (d_{0k} - \bar{d}_0)(\bar{x}_k - \bar{x})}{\sum n_k (\bar{x}_k - \bar{x})^2} \tag{B8}$$

where $\bar{d}_0 = \sum n_k d_{0k}/n$, and

$$b_0 = \bar{d}_0 - b_2 \bar{x} \tag{B9}$$

For these weighted estimators the weighted sum of the resulting residuals equals zero, that is,

$$\sum n_k u_k = 0 \tag{B10}$$

instead of the unweighted residuals in Eq. (B6). This property of the u's is used in several other appendixes.

Similarly,

$$b_3 = \frac{\sum n_k (d_{1k} - \bar{d}_1)(\bar{x}_k - \bar{x})}{\sum n_k (\bar{x}_k - \bar{x})^2} \tag{B11}$$

where $d_1 = \sum n_k d_{1k}/n$, and

$$b_1 = \bar{d}_1 - b_3 \bar{x} \tag{B12}$$

The v residuals obtained from these estimators satisfy the equation

$$\sum n_k \bar{x}_k v_k = 0 \tag{B13}$$

instead of the similar equation for the unweighted residuals in Eq. (B7). This property of the v's is used in several other appendixes.

The same reasoning applies to the centered model.

Appendix

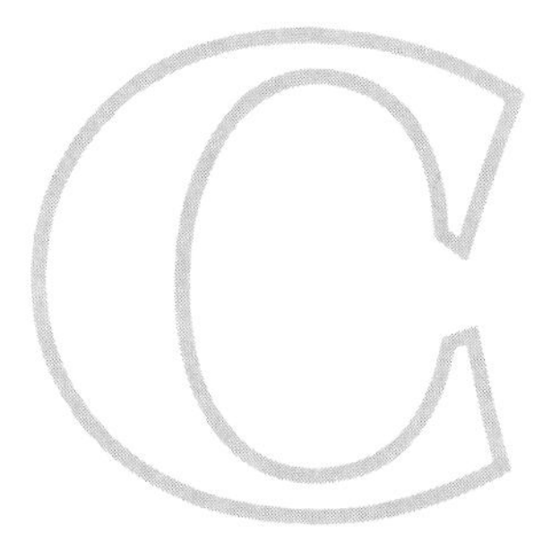

Comparing Estimates from Single and Separate Equations

Residuals from the single and separate equation procedures, as in Chapter 3, are related according to the equation

$$e_{ik} = u_k + v_k x_{ik} + f_{ik} \tag{C1}$$

If the e's are obtained directly from the single multiple-regression equation, they can be found from the equation

$$e_{ik} = y_{ik} - b_0 - b_1 x_{ik} - b_2 \bar{x}_k - b_3 x_{ik} \bar{x}_k \tag{C2}$$

By definition, the sum of these residuals equals zero.

On the other hand, from Eq. (C1),

$$\begin{aligned} \sum\sum e_{ik} &= \sum\sum u_k + \sum\sum v_k x_{ik} + \sum\sum f_{ik} \\ &= \sum n_k u_k + \sum n_k v_k \bar{x}_k + 0 \end{aligned} \tag{C3}$$

The sum of the f's equals zero within each group. If the number of observations is the same in each group, or if n_k is used as the weight for the kth group, we know from Appendix B that the first two sums in the equation above are also equal to zero.

Further conditions that need to be satisfied for the b values from the single and the separate equations to be the same can be obtained by considering the normal equations used to solve for the b's. For example, the b's in Eq. (C2) are the solutions to a set of four normal equations.

The two coefficients d_{0k} and d_{1k} are obtained from the two normal equations

$$d_{0k} n_k + d_{1k} \sum x_{ik} = \sum y_{ik} \tag{C4}$$

$$d_{0k} \sum x_{ik} + d_{1k} \sum x_{ik}^2 = \sum x_{ik} y_{ik} \tag{C5}$$

According to the basic model, d_{0k} and d_{1k} are functions of $\bar{X}$. Substituting d_{0k} and d_{1k} the two equations become

$$(b_0 + b_2\bar{x}_k + u_k)n_k + (b_1 + b_3\bar{x}_k + v_k)\sum x_{ik} = \sum y_{ik} \tag{C6}$$

$$(b_0 + b_2\bar{x}_k + u_k)\sum x_{ik} + (b_1 + b_3\bar{x}_k + v_k)\sum x_{ik}^2 = \sum x_{ik}y_{ik} \tag{C7}$$

By adding across the groups these equations can be written

$$\begin{aligned} &b_0\sum n_k + b_1\sum\sum x_{ik} + b_2\sum n_k\bar{x}_k + b_3\sum\sum x_{ik}\bar{x}_k \\ &\quad + \left(\sum n_k u_k + \sum\sum x_{ik}v_k\right) = \sum\sum y_{ik} \end{aligned} \tag{C8}$$

$$\begin{aligned} &b_0\sum\sum x_{ik} + b_1\sum\sum x_{ik}^2 + b_2\sum\sum x_{ik}\bar{x}_k + b_3\sum\sum x_{ik}^2\bar{x}_k \\ &\quad + \left(\sum\sum x_{ik}u_k + \sum\sum x_{ik}^2 v_k\right) = \sum\sum x_{ik}y_{ik} \end{aligned} \tag{C9}$$

Now we look for the conditions under which the terms in the parentheses equal zero. Using n_k as weights we know from Appendix B that

$$\sum n_k u_k = 0 \tag{C10}$$

$$\sum\sum v_k x_{ik} = \sum v_k n_k \bar{x}_k = 0 \tag{C11}$$

$$\sum\sum u_k x_{ik} = \sum u_k n_k \bar{x}_k = 0 \tag{C12}$$

Finally, the last sum can be written

$$\sum\sum v_k x_{ik}^2 = \sum n_k v_k s_k^2 + \sum n_k v_k \bar{x}_k^2 \tag{13}$$

where the variance of X in the kth group is denoted by s_k^2. When this variance has the same value in each group, the first sum on the right-hand side of Eq. (C13) becomes

$$\sum n_k v_k s_k^2 = s^2\sum n_k v_k = 0 \tag{C14}$$

When the variances are not equal, this sum equals zero if v_k and s_k^2 are uncorrelated. This means that group homogeneity should not influence the slopes for the relationship between Y and X in the groups. If it is thought that group homogeneity does affect the relationship, we may want to consider this factor explicitly and work with a model of the form

$$d_{0k} = b_0 + b_2\bar{x}_k + b_4 s_k^2 + u_k \tag{C15}$$

$$d_{1k} = b_1 + b_3\bar{x}_k + b_5 s_k^2 + v_k \tag{C16}$$

When v_k is uncorrelated with $\bar{x}_k^2$, the second sum on the right-hand side of Eq. (C13) equals zero. This occurs when the within-group slopes are linearly related to the group means, so that no term for $\bar{x}_k^2$ belongs in the model.

Additional conditions that have to be satisfied for the single and the separate equations to give the same values for the coefficients are obtained by multiplying

Eqs. (C6) and (C7) by $\bar{x}_k$ and adding across the groups. That gives the equations

$$b_0\sum n_k\bar{x}_k + b_1\sum\sum x_{ik}\bar{x}_k + b_2\sum n_k\bar{x}_k^2 + b_3\sum\sum x_{ik}\bar{x}_k^2 + (\sum n_k u_k\bar{x}_k + \sum\sum x_{ik}v_k\bar{x}_k) = \sum\sum y_{ik}\bar{x}_k \quad \text{(C17)}$$

$$b_0\sum\sum x_{ik}\bar{x}_k + b_1\sum\sum x_{ik}^2\bar{x}_k + b_2\sum\sum x_{ik}\bar{x}_k^2 + b_3\sum\sum x_{ik}^2\bar{x}_k^2 + (\sum\sum x_{ik}u_k\bar{x}_k + \sum\sum v_k x_{ik}^2\bar{x}_k) = \sum\sum x_{ik}\, y_{ik}\bar{x}_k \quad \text{(C18)}$$

For the sums in the parentheses,

$$\sum n_k u_k \bar{x}_k = 0 \quad \text{(C19)}$$

as long as n_k is used as weight for the kth group. Similarly,

$$\sum\sum v_k x_{ik}\bar{x}_k = \sum n_k v_k \bar{x}_k^2 = 0 \quad \text{(C20)}$$

provided that d_{1k} is linearly related to the group mean. Also,

$$\sum\sum u_k x_{ik}\bar{x}_k = \sum n_k u_k \bar{x}_k^2 = 0 \quad \text{(C21)}$$

when d_{0k} is linearly related to the group mean. Finally,

$$\sum\sum v_k x_{ik}^2\bar{x}_k = \sum n_k v_k \bar{x}_k s_k^2 + \sum n_k v_k \bar{x}_k^3 \quad \text{(C22)}$$

The last sum equals zero as long as v_k is uncorrelated with $\bar{x}_k^3$, which is the case when the relationship between d_{1k} and the group mean is linear. The product $\bar{x}_k s_k^2$ can be interpreted as the interaction between the group level and group homogeneity of X. When v_k, and thereby d_{1k}, is uncorrelated with this interaction term, the corresponding sum in Eq. (C22) will equal zero.

When the terms in the parentheses in Eqs. (C8), (C9), (C17), and (C18) equal zero, what remains in those four equations forms the set of four normal equations for the computation of the b's from the single equation. Since the b's do satisfy those equations, the separate and single equations will give the same estimates as long as the conditions in this appendix hold.

In summary, using n_k as the weight for the separate equations, the conditions derived here require that the group intercepts and slopes be only linearly related to the group means and not related to the within-group variances of the explanatory variable.

Appendix

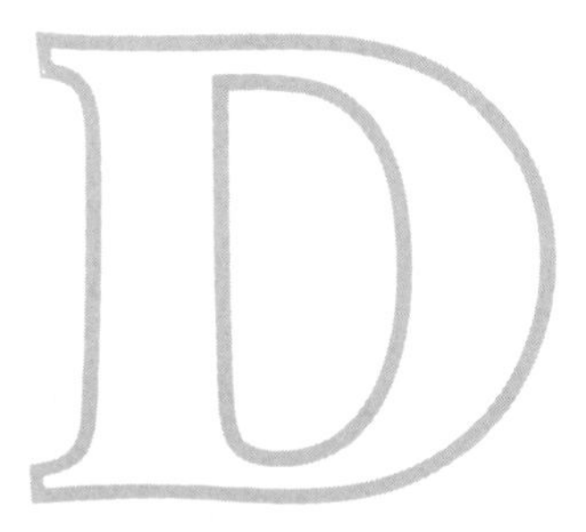

Partitioning Sums of Squares with Centering

For the centered model the variables are related according to the equation

$$y'_{ik} = a_0 + a_1(x_{ik} - \bar{x}_k) + a_2(\bar{x}_k - \bar{x}) + a_3(x_{ik} - \bar{x}_k)(\bar{x}_k - \bar{x}) + e_{ik} \qquad \text{(D1)}$$

If the three explanatory variables are uncorrelated among themselves, the regression sum of squares can be decomposed into three unique parts, one for each variable.

All three variables have means equal to zero. To show that the variables are uncorrelated we therefore need to show that their sums of cross products (scp) equal zero. Here

$$\begin{aligned} \text{scp(individual and group)} &= \sum\sum(x_{ik} - \bar{x}_k)(\bar{x}_k - \bar{x}) \\ &= \sum_k (\bar{x}_k - \bar{x}) \sum_i (x_{ik} - \bar{x}_k) = 0 \end{aligned} \qquad \text{(D2)}$$

since the sum of the deviations around the mean equals zero for each group. For the same reason,

$$\begin{aligned} \text{scp(group and interaction)} &= \sum\sum(\bar{x}_k - \bar{x})\{(x_{ik} - \bar{x}_k)(\bar{x}_k - \bar{x})\} \\ &= \sum_k (\bar{x}_k - \bar{x})^2 \sum_i (x_{ik} - \bar{x}_k) = 0 \end{aligned} \qquad \text{(D3)}$$

Finally,

$$\begin{aligned} \text{scp(individual and interaction)} &= \sum\sum(x_{ik} - \bar{x}_k)\{(x_{ik} - \bar{x}_k)(\bar{x}_k - \bar{x})\} \\ &= \sum_k (\bar{x}_k - \bar{x}) \sum_i (x_{ik} - \bar{x}_k)^2 = \sum_k (\bar{x}_k - \bar{x}) n_k s_k^2 \end{aligned} \qquad \text{(D4)}$$

where s_k^2 is the variance of X in the kth group. The sum in Eq. (D4) will equal zero when the within-group variances are equal. More generally, the sum will equal zero when the within-group variances are uncorrelated with the group means. When the three explanatory variables are uncorrelated, the regression sum of squares can be partitioned in standard ways into three components, one for each variable.

The unexplained variation can also be partitioned into three components. We get

$$\begin{aligned}\sum\sum e_{ik}^2 &= \sum\sum(u_k + v_k(x_{ik} - \bar{x}_k) + f_{ik})^2 \\ &= \sum n_k u_k^2 + \sum\sum v_k^2(x_{ik} - \bar{x}_k)^2 + \sum\sum f_{ik}^2\end{aligned} \tag{D5}$$

because the three corresponding sums of cross products all equal zero. This can be seen from

$$\sum\sum u_k v_k(x_{ik} - \bar{x}_k) = \sum_k u_k v_k \sum_i (x_{ik} - \bar{x}_k) = 0 \tag{D6}$$

since the sum of deviations around the group mean equals zero within each group. Also,

$$\sum\sum u_k f_{ik} = \sum_k u_k \sum_i f_{ik} = 0 \tag{D7}$$

since the sum of the residuals within each group equals zero. Finally,

$$\sum\sum v_k(x_{ik} - \bar{x}_k)f_{ik} = \sum_k v_k \sum_i (x_{ik} - \bar{x}_k)f_{ik} = 0 \tag{D8}$$

since the residuals are uncorrelated with the explanatory variable within each group.

Appendix

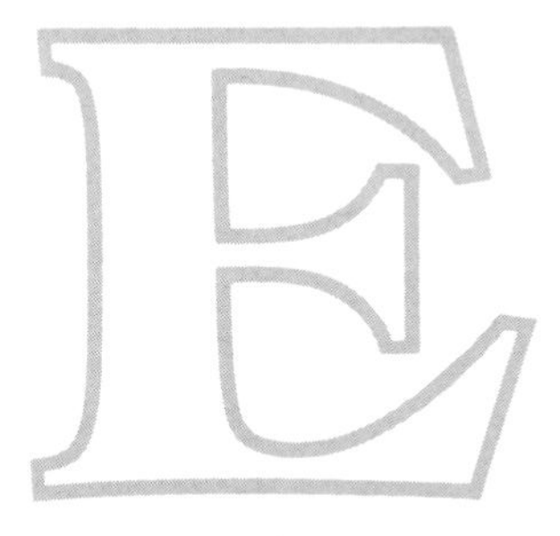

Contextual Analysis Using Contingency Tables with Many Observations

For a set of contingency tables the single equation is

$$y_{ik} = b_0 + b_1 x_{ik} + b_2 p_{\cdot 2k} + b_3 x_{ik} p_{\cdot 2k} + e_{ik} \tag{E1}$$

where Y and X are dummy variables. Assigning 0's and 1's for the dummy variables can be cumbersome if there are many observations in each table, for example, if the tables consist of census data for millions of people in different states.

Instead of entering the raw data in the computer for the computations of regression coefficients and sums of squares, it is easier to input the means, standard deviations, and correlation coefficients for the variables. The formulas for these quantities are given in this appendix.

The following notation is used for the data in the kth group as displayed in 2×2 tables:

FREQUENCIES

		X: 0	X: 1	
Y	1	n_{11k}	n_{12k}	$n_{1\cdot k}$
	0	n_{21k}	n_{22k}	$n_{2\cdot k}$
		$n_{\cdot 1k}$	$n_{\cdot 2k}$	n_k

PROPORTIONS

		X: 0	X: 1	
Y	1	p_{11k}	p_{12k}	$p_{1\cdot k}$
	0	p_{21k}	p_{22k}	$p_{2\cdot k}$
		$p_{\cdot 1k}$	$p_{\cdot 2k}$	1.00

Adding across groups, $n_{\cdot 2} = \sum n_{\cdot 2k}$, $n_{1\cdot} = \sum n_{1\cdot k}$, and $n = \sum n_k$.

Means

Y: $\quad \bar{y} = n_{1\cdot}/n = p_{1\cdot}$ (E2)

X: $\quad \bar{x} = n_{\cdot 2}/n = p_{\cdot 2}$ (E3)

$$P_{\cdot 2}: \qquad \bar{p}_{\cdot 2} = \sum n_k p_{\cdot 2k} / \sum n_k = \sum n_{\cdot 2k}/n = p_{\cdot 2} \tag{E4}$$

$$XP_{\cdot 2}: \qquad \overline{xp}_{\cdot 2} = \sum\sum x_{ik} p_{\cdot 2k}/n = \sum n_{\cdot 2k} p_{\cdot 2k}/n \tag{E5}$$

Standard Deviations

For a variable W the *variance* can be found from

$$s_w^2 = \frac{\sum(w - \bar{w})^2}{n} = \frac{\sum w^2 - (\sum w)^2/n}{n} = \frac{\sum w^2}{n} - \bar{w}^2 \tag{E6}$$

Using this formula the variances for the four variables are shown below. The *standard deviations* are found by taking the square roots of the variances.

$$Y: \qquad \sum\sum y_{ik}^2 = \sum\sum y_{ik} = \sum n_{1 \cdot k} = n_{1 \cdot}$$

$$s_y^2 = \frac{n_{1 \cdot}}{n} - p_{1 \cdot}^2 = p_{1 \cdot}(1 - p_{1 \cdot}) = p_{1 \cdot} p_{2 \cdot} \tag{E7}$$

$$X: \qquad s_x^2 = p_{\cdot 1} p_{\cdot 2} \tag{E8}$$

$$P_{\cdot 2}: \qquad \sum\sum p_{\cdot 2k}^2 = \sum n_k p_{\cdot 2k}^2 = \sum n_{\cdot 2k} p_{\cdot 2k}$$

$$s_{p \cdot 2}^2 = \frac{\sum n_{\cdot 2k} p_{\cdot 2k}}{n} - p_{\cdot 2}^2 \tag{E9}$$

$$XP_{\cdot 2}: \qquad \sum\sum (x_{ik} p_{\cdot 2k})^2 = \sum\sum x_{ik} p_{\cdot 2k}^2 = \sum n_{\cdot 2k} p_{\cdot 2k}^2$$

$$s_{xp \cdot 2}^2 = \frac{\sum n_{\cdot 2k} p_{\cdot 2k}^2}{n} - \left(\frac{\sum n_{\cdot 2k} p_{\cdot 2k}}{n} \right)^2 \tag{E10}$$

Correlation Coefficients

For any two variables X_1 and X_2 the correlation coefficient r_{12} can be found from the expression

$$r_{12} = \frac{\sum x_1 x_2/n - \bar{x}_1 \bar{x}_2}{s_1 s_2} \tag{E11}$$

Using this formula to find the correlation coefficients gives

$$Y \text{ and } X: \qquad \sum\sum x_{ik} y_{ik} = \sum n_{12k} = n_{12}$$

$$r_{xy} = \frac{n_{12}/n - p_{\cdot 2} p_{1 \cdot}}{s_x s_y} = \frac{p_{12} - p_{1\cdot} p_{\cdot 2}}{s_x s_y} \tag{E12}$$

Y and $P_{.2}$:

$$\sum\sum y_{ik}p_{\cdot 2k} = \sum n_{1\cdot k}p_{\cdot 2k}$$

$$r_{yp_{\cdot 2}} = \frac{\sum n_{1\cdot k}p_{\cdot 2k}/n - p_{1\cdot}p_{\cdot 2}}{s_y s_{p_{\cdot 2}}} \tag{E13}$$

Y and $XP_{.2}$:

$$\sum\sum y_{ik}x_{ik}p_{\cdot 2k} = \sum n_{12k}p_{\cdot 2k}$$

$$r_{y(xp_{\cdot 2})} = \frac{\sum n_{12k}p_{\cdot 2k}/n - p_{1\cdot}(\sum n_{\cdot 2k}p_{\cdot 2k}/n)}{s_y s_{xp_{\cdot 2}}} \tag{E14}$$

X and $P_{.2}$:

$$\sum\sum x_{ik}p_{\cdot 2k} = \sum n_{\cdot 2k}p_{\cdot 2k}$$

$$r_{xp_{\cdot 2}} = \frac{\sum n_{\cdot 2k}p_{\cdot 2k}/n - p_{\cdot 2}^2}{s_x s_{p_{\cdot 2}}} \tag{E15}$$

X and $XP_{.2}$:

$$\sum\sum x_{ik}x_{ik}p_{\cdot 2k} = \sum n_{\cdot 2k}p_{\cdot 2k}$$

$$r_{x(xp_{\cdot 2})} = \frac{\sum n_{\cdot 2k}p_{\cdot 2k}/n - p_{\cdot 2}\sum n_{\cdot 2k}p_{\cdot 2k}/n}{s_x s_{xp_{\cdot 2}}}$$

$$= \frac{\overline{xp}_{\cdot 2} - p_{\cdot 2}\overline{xp}_{\cdot 2}}{s_x s_{p_{\cdot 2}}}$$

$$= \frac{p_{\cdot 1}\overline{xp}_{\cdot 2}}{s_x s_{p_{\cdot 2}}} \tag{E16}$$

$P_{.2}$ and $XP_{.2}$:

$$\sum\sum p_{\cdot 2k}x_{ik}p_{\cdot 2k} = \sum n_{\cdot 2k}p_{\cdot 2k}^2$$

$$r_{p_{\cdot 2}(xp_{\cdot 2})} = \frac{\sum n_{\cdot 2k}p_{\cdot 2k}^2/n - p_{\cdot 2}\sum n_{\cdot 2k}p_{\cdot 2k}/n}{s_{p_{\cdot 2}}s_{xp_{\cdot 2}}} \tag{E17}$$

Appendix

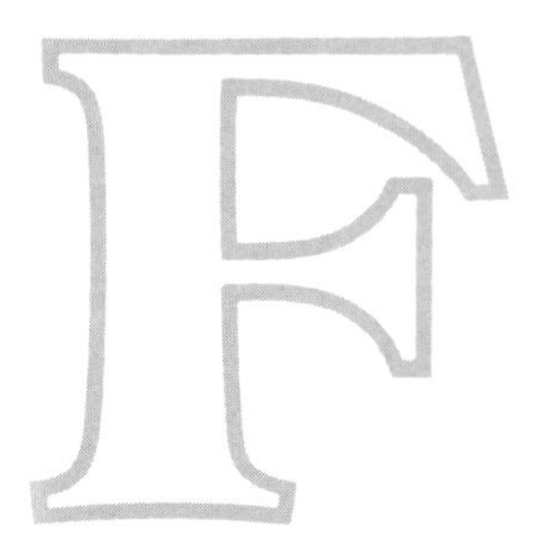

Impact of Nonlinear Context Effects on the Regression Coefficients

The within-group regression coefficients may be related to the group means according to a model containing a quadratic term, as seen in the equations

$$d_{0k} = b_0 + b_2\bar{x}_k + b_4\bar{x}_k^2 + u_k \tag{F1}$$

$$d_{1k} = b_1 + b_3\bar{x}_k + b_5\bar{x}_k^2 + v_k \tag{F2}$$

If the quadratic term is not included and the data are analyzed according to the linear model,

$$d_{0k} = b_0 + b_2\bar{x}_k + u'_k \tag{F3}$$

$$d_{1k} = b_1 + b_3\bar{x}_k + v'_k \tag{F4}$$

then the various residuals are related according to the equations

$$u'_k = b_4\bar{x}_k^2 + u_k \tag{F5}$$

$$v'_k = b_5\bar{x}_k^2 + v_k \tag{F6}$$

In this case the residuals from the linear model would be correlated with $\bar{X}^2$. From Appendix C we know that in such a case the estimates from the single equation and the two separate equations will not coincide. This would alert us that the model is incorrectly specified, and perhaps a square term should be included.

This process can be generalized. When the $\bar{x}_k^2$ term is included in the model, the single and the two separate equations give the same estimates as long as the residuals u_k and v_k are uncorrelated with the within-group variances s_k^2, the third-degree terms $\bar{x}_k^3$, the fourth-degree terms $\bar{x}_k^4$, and the two interaction terms $\bar{x}_k s_k^2$ and $\bar{x}_k^2 s_k^2$. In short, when the two procedures result in approximately the same estimates, we have an indication that the model is correctly specified and that terms like the ones mentioned above should not explicitly be brought into the model.

These conditions are arrived at the same way as the similar conditions in Appendix C are obtained. The two separate Eqs. (F1) and (F2) result in the following single equation relating Y and the full set of variables derived from X:

$$y_{ik} = b_0 + b_1 x_{ik} + b_2 \bar{x}_k + b_3 x_{ik}\bar{x}_k + b_4 \bar{x}_k^2 + b_5 x_{ik}\bar{x}_k^2 + e_{ik} \tag{F7}$$

The coefficients in this equation are obtained by solving the six normal equations that come from minimizing the residual sum of squares.

On the other hand, the first of these normal equations can also be arrived at by substituting d_{0k} and d_{1k} from Eqs. (F1) and (F2) into Eq. (C4), which is one of the normal equations for d_{0k} and d_{1k}, adding over k, and requiring

$$\sum n_k u_k + \sum n_k v_k \bar{x}_k = 0 \tag{F8}$$

These two sums both add to zero if n_k is used as a weight for the kth group, since the weighted sum of the residuals will equal zero and the residual variable V is uncorrelated with $\bar{X}$.

The second normal equation for the b's is obtained by substituting d_{0k} and d_{1k} from Eqs. (F1) and (F2) into Eq. (C5), adding over k, and requiring

$$\sum\sum u_k x_{ik} + \sum\sum v_k x_{ik}^2 = \sum n_k u_k \bar{x}_k + \sum n_k v_k s_k^2 + \sum n_k v_k \bar{x}_k^2 = 0 \tag{F9}$$

The first and third sums on the right equal zero because $\bar{X}$ and $\bar{X}^2$ are both variables in the model, and the residuals are therefore uncorrelated with these variables. The middle sum equals zero when the within-group variances of X are all equal or when the v's, and thereby the within-group slopes, are uncorrelated with the variances s_k^2. Thus the group homogeneity should not affect the slope of the relationship between Y and X within the groups.

The third normal equation for the b's is obtained by substituting d_{0k} and d_{1k} into Eq. (C4), multiplying both sides by $\bar{x}_k$, and adding over k. The resulting equation requires

$$\sum n_k u_k \bar{x}_k + \sum\sum v_k x_{ik}\bar{x}_k = \sum n_k u_k \bar{x}_k + \sum n_k v_k \bar{x}_k^2 = 0 \tag{F10}$$

Both sums equal zero, since $\bar{x}_k$ and $\bar{x}_k^2$ are terms in the model, and the residuals are uncorrelated with these variables.

The fourth normal equation is obtained by substituting d_{0k} and d_{1k} into Eq. (C5), multiplying by $\bar{x}_k$, and adding over k. The fifth and sixth normal equations are obtained by substituting d_{0k} and d_{1k} into Eqs. (C4) and (C5), multiplying by $\bar{x}_k^2$, and adding over k. For these equations to be identical to the corresponding normal equations for the b's from the single equation, and thereby show that the b values from the single equation are equal to those from the two separate equations, the following conditions must be satisfied:

$$\sum n_k u_k \bar{x}_k^3 = 0 \tag{F11}$$

$$\sum n_k v_k \bar{x}_k^3 = 0 \tag{F12}$$

$$\sum n_k v_k \bar{x}_k^4 = 0 \tag{F13}$$

$$\sum n_k v_k \bar{x}_k s_k^2 = 0 \tag{F14}$$

$$\sum n_k v_k \bar{x}_k^2 s_k^2 = 0 \tag{F15}$$

To summarize, for a model containing the $\bar{x}_k^2$ term in the two separate equations the two sets of b values will be equal when the within-group intercepts are not correlated with $\bar{x}_k^3$ and the slopes are not correlated with $\bar{x}_k^3$, $\bar{x}_k^4$, s_k^2, $\bar{x}_k s_k^2$, and $\bar{x}_k^2 s_k^2$.

Appendix

Decomposing Variation with Several Sets of Explanatory Variables

The decomposition of explained and unexplained variation with several sets of explanatory variables follows the pattern for one variable.

The uncentered model permits the decomposition of the unexplained variation into one component due to the individual-level variables and one component due to the group and interaction variables. As an example and without any loss of generality, for two explanatory variables the residuals are related as in Eq. (6.16)

$$e_{ik} = f_{ik} + u_{0k} + u_{1k}x_{1ik} + u_{2k}x_{2ik} \tag{G1}$$

The following sums of cross products equal zero:

$$\sum\sum f_{ik}u_{0k} = \sum_k u_{0k} \sum_i f_{ik} = 0 \tag{G2}$$

since the residuals add to zero within each group;

$$\sum\sum f_{ik}u_{1k}x_{1ik} = \sum_k u_{1k} \sum_i f_{ik}x_{1ik} = 0 \tag{G3}$$

$$\sum\sum f_{ik}u_{2k}x_{2ik} = \sum_k u_{2k} \sum_i f_{ik}x_{2ik} = 0 \tag{G4}$$

since the residuals are uncorrelated with the explanatory variables within each group.

By squaring both sides of Eq. (G1), adding, and using the results expressed in Eqs. (G2), (G3), and (G4), we get

$$\sum\sum e_{ik}^2 = \sum\sum f_{ik}^2 + \text{(sums involving group and interaction terms)} \tag{G5}$$

This means that the total residual variation can be decomposed into one part representing the individual-level unexplained variation and another representing the group and interaction-level unexplained variation.

The centered model permits a more detailed decomposition because it has more uncorrelated variables. The explained variation can be decomposed into three parts, one for each level of variables, subject to the following considerations. Each individual-level variable is uncorrelated with each of the group variables, for example,

$$\sum\sum(x_{1ik} - \bar{x}_{1k})(\bar{x}_{2k} - \bar{x}_2) = \sum_k (\bar{x}_{2k} - \bar{x}_2) \sum_i (x_{1ik} - \bar{x}_{1k}) = 0 \qquad \text{(G6)}$$

because the sum of deviations around the group mean equals zero within each group. Each group variable is uncorrelated with all the interaction variables for the same reason.

The individual and interaction variables are sometimes related. We get

$$\sum\sum(x_{rik} - \bar{x}_{rk})\{(x_{rik} - \bar{x}_{rk})(\bar{x}_{sk} - \bar{x}_s)\} = \sum(\bar{x}_{sk} - \bar{x}_s)n_k s_{rk}^2 \qquad \text{(G7)}$$

and these two variables will be uncorrelated when the within-group variances of X_r are the same in each of the groups. Similarly,

$$\begin{aligned}\sum\sum(x_{rik} - \bar{x}_{rk})\{(x_{sik} - \bar{x}_{sk})(\bar{x}_{tk} - \bar{x}_t)\} &= \sum_k (\bar{x}_{tk} - \bar{x}_t) \sum_i (x_{rik} - \bar{x}_{rk})(x_{sik} - \bar{x}_{sk}) \\ &= \sum_k (\bar{x}_{tk} - \bar{x}_t) n_k \operatorname{cov}(x_{rk}, x_{sk}) \qquad \text{(G8)}\end{aligned}$$

where the covariance term refers to the covariance of X_r and X_s in the kth group. This sum will equal zero when these covariances are equal in all the groups. If the variables have constant variances across the groups, the requirement of equal covariances translates into a requirement of constant correlation coefficients between the two variables in all the groups.

Thus for a centered model it is possible to decompose the explained variation into one part due to the group-level variables and one part due to the individual and interaction variables. Furthermore, if the explanatory variables have constant variances and covariances across the groups, then separate components for the individual and interaction variables can be obtained.

The unexplained variation in the centered model can be decomposed into three parts for the same reasons as explained in Appendix D for one set of explanatory variables.

Appendix

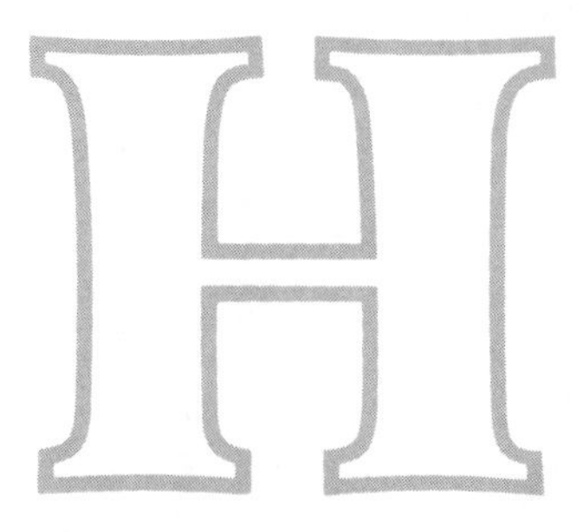

Equations Relating Various Regression Coefficients and η^2

Let η^2 equal the ratio of the between-group sum of squares and the total sum of squares for X, that is,

$$\eta^2 = \frac{\sum n_k(\bar{x}_k - \bar{x})^2}{\sum\sum(x_{ik} - \bar{x})^2}, \qquad 1 - \eta^2 = \frac{\sum\sum(x_{ik} - \bar{x}_k)^2}{\sum\sum(x_{ik} - \bar{x})^2} \tag{H1}$$

The regression coefficient d_{1p} is obtained when Y is regressed on X for all the data, where

$$d_{1p} = \frac{\sum\sum(x_{ik} - \bar{x})(y_{ik} - \bar{y})}{\sum\sum(x_{ik} - \bar{x})^2} \tag{H2}$$

By partitioning the total covariation in the numerator into the within-group and between-group covariation, d_{1p} can be written

$$d_{1p} = \frac{\sum\sum(x_{ik} - \bar{x}_k)(y_{ik} - \bar{y}_k) + \sum n_k(\bar{x}_k - \bar{x})(\bar{y}_k - \bar{y})}{\sum\sum(x_{ik} - \bar{x})^2} \tag{H3}$$

Regressing Y on X in the kth group results in the regression coefficient

$$d_{1k} = \frac{\sum(x_{ik} - \bar{x}_k)(y_{ik} - \bar{y}_k)}{\sum(x_{ik} - \bar{x}_k)^2} \tag{H4}$$

and thereby

$$\sum_i (x_{ik} - \bar{x}_k)(y_{ik} - \bar{y}_k) = d_{1k} \sum_i (x_{ik} - \bar{x}_k)^2 \tag{H5}$$

Similarly, regressing the group means $\bar{Y}$ on $\bar{X}$ results in the coefficient

$$d_{1g} = \frac{\sum n_k(\bar{x}_k - \bar{x})(\bar{y}_k - \bar{y})}{\sum n_k(\bar{x}_k - \bar{x})^2} \tag{H6}$$

and thereby

$$\sum n_k(\bar{x}_k - \bar{x})(\bar{y}_k - \bar{y}) = d_{1g}\sum n_k(\bar{x}_k - \bar{x})^2 \tag{H7}$$

Substituting from Eqs. (H5) and (H7) into Eq. (H3) gives

$$d_{1p} = \frac{\sum d_{1k}\sum(x_{ik} - \bar{x}_k)^2 + d_{1g}\sum n_k(\bar{x}_k - \bar{x})^2}{\sum\sum(x_{ik} - \bar{x})^2} \tag{H8}$$

According to the model, d_{1k} is related to $\bar{x}_k$ as seen in the equation

$$d_{1k} = b_1 + b_3\bar{x}_k + v_k \tag{H9}$$

Substituting this expression for d_{1k} into Eq. (H8) gives

$$\begin{aligned} d_{1p} &= \frac{\sum(b_1 + b_3\bar{x}_k + v_k)\sum(x_{ik} - \bar{x}_k)^2 + d_{1g}\sum n_k(\bar{x}_k - \bar{x})^2}{\sum\sum(x_{ik} - \bar{x})^2} \\ &= b_1(1 - \eta^2) + b_3\frac{\sum\bar{x}_k\sum(x_{ik} - \bar{x}_k)^2}{\sum\sum(x_{ik} - \bar{x})^2} + \frac{\sum v_k\sum(x_{ik} - \bar{x}_k)^2}{\sum\sum(x_{ik} - \bar{x})^2} + d_{1g}\eta^2 \end{aligned} \tag{H10}$$

If the variance of X is the same in each group and equal to s^2, then the first numerator on the right-hand side in Eq. (H10) can be written

$$\begin{aligned} \sum\bar{x}_k\sum(x_{ik} - \bar{x}_k)^2 &= \sum\bar{x}_k n_k s^2 = s^2\bar{x}n = \bar{x}s^2\sum n_k = \bar{x}\sum s^2 n_k \\ &= \bar{x}\sum_k \frac{\sum(x_{ik} - \bar{x}_k)^2}{n_k} n_k = \bar{x}\sum\sum(x_{ik} - \bar{x}_k)^2 \end{aligned} \tag{H11}$$

This result could also have been arrived at by regarding the common variance s^2 as the pooled estimate of the common variance across the groups. The second numerator becomes

$$\sum v_k\sum(x_{ik} - \bar{x}_k)^2 = \sum v_k n_k s^2 = s^2\sum n_k v_k = 0 \tag{H12}$$

Substituting the results in Eqs. (H11) and (H12) back into Eq. (H10) gives

$$\begin{aligned} d_{1p} &= b_1(1 - \eta^2) + b_3\bar{x}\frac{\sum\sum(x_{ik} - \bar{x}_k)^2}{\sum\sum(x_{ik} - \bar{x})^2} + d_{1g}\eta^2 \\ &= b_1(1 - \eta^2) + b_3\bar{x}(1 - \eta^2) + d_{1g}\eta^2 \end{aligned} \tag{H13}$$

This is Eq. (8.29).

Various special cases of this equation are used in Chapter 8. When there is no interaction present in the data, the coefficient b_3 equals zero and the equation then reduces to

$$d_{1p} = b_1(1 - \eta^2) + d_{1g}\eta^2 \tag{H14}$$

This is Eq. (8.16).

If the group means are not related, the coefficient d_{1g} will equal zero. In that case Eq. (H13) reduces to

$$d_{1p} = b_1(1 - \eta^2) + b_3\bar{x}(1 - \eta^2) \tag{H15}$$

When there is no interaction effect present in the data, $\bar{Y}$ and $\bar{X}$ are related such that

$$d_{1g} = b_1 + b_2 \tag{H16}$$

as shown in Eq. (8.12).

Substituting this sum for d_{1g} into Eq. (H14) results in

$$d_{1p} = b_1(1 - \eta^2) + (b_1 + b_2)\eta^2 = b_1 + b_2\eta^2 \tag{H17}$$

This is Eq. (8.15). Finally, substituting b_1 from Eq. (H16) into Eq. (H14) gives

$$d_{1p} = (d_{1g} - b_2)(1 - \eta^2) + d_{1g}\eta^2 = d_{1g} - b_2(1 - \eta^2) \tag{H18}$$

This is Eq. (8.17).

Appendix

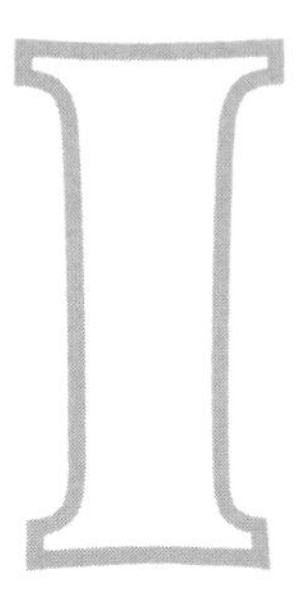

Multilevel Analysis with Incomplete Data, Using Correlations as Inputs into Regression Analyses

Standard regression programs can use means, standard deviations, and correlation coefficients instead of raw data as input for the analysis. Many of these quantities can be obtained directly from group data, and a few require individual-level data. Sometimes it is possible to use data from several sources to find the quantities required for matrix input.

The model is of the form

$$y_{ik} = b_0 + b_1 x_{ik} + b_2 \bar{x}_k + b_3 x_{ik}\bar{x}_k + e_{ik} \tag{I1}$$

and the analysis requires the means, standard deviations, and correlation coefficients for the variables Y, X, $\bar{X}$, and $X\bar{X}$.

Means

$$Y: \quad \bar{y} = \sum\sum y_{ik}/n = \sum n_k \bar{y}_k/n \tag{I2}$$

$$X: \quad \bar{x} = \sum\sum x_{ik}/n = \sum n_k \bar{x}_k/n \tag{I3}$$

$$\bar{X}: \quad \bar{\bar{x}} = \sum n_k \bar{x}_k/n = \bar{x} \tag{I4}$$

$$X\bar{X}: \quad \overline{x\bar{x}} = \sum\sum x_{ik}\bar{x}_k/n = \sum n_k \bar{x}_k^2/n \tag{I5}$$

Standard Deviations

The standard deviations are found as the square roots of the following *variances*:

$$Y: \quad s_y^2 = \frac{\sum\sum y_{ik}^2}{n} - \bar{y}^2 = \frac{\sum n_k s_{yk}^2 + \sum n_k \bar{y}_k^2}{n} - \bar{y}^2 \tag{I6}$$

$$X: \quad s_x^2 = \frac{\sum\sum x_{ik}^2}{n} - \bar{x}^2 = \frac{\sum n_k s_{xk}^2 + \sum n_k \bar{x}_k^2}{n} - \bar{x}^2 \tag{I7}$$

$$\bar{X}: \quad s_{\bar{x}}^2 = \frac{\sum n_k(\bar{x}_k - \bar{x})^2}{n} = \frac{\sum n_k \bar{x}_k^2}{n} - \bar{x}^2 \tag{I8}$$

$$X\bar{X}: \quad s_{x\bar{x}}^2 = \frac{\sum\sum(x_{ik}\bar{x}_k)^2}{n} - (\overline{x\bar{x}})^2 = \frac{\sum n_k x_k s_{xk}^2 + \sum n_k \bar{x}_k^2}{n} - (\overline{x\bar{x}})^2 \tag{I9}$$

These variances are found directly from the marginal distributions of X and Y or from the within-group means and variances of X and Y.

Correlation Coefficients

Y and X: The sum $\sum\sum x_{ik} y_{ik}$ requires the joint distribution of X and Y on the individual level. Without individual-level data it may be possible to find r_{xy} directly from some sample survey of the same population.

$$r_{xy} = \frac{(\sum\sum(x_{ik}y_{ik})/n) - \bar{x}\bar{y}}{s_x s^y} \tag{I10}$$

Y and $\bar{X}$: $\sum\sum y_{ik}\bar{x}_k = \sum n_k \bar{x}_k \bar{y}_k$

$$r_{\bar{x}y} = \frac{(\sum(n_k \bar{x}_k \bar{y}_k)/n) - \bar{x}\bar{y}}{s_{\bar{x}} s_y} \tag{I11}$$

Y and $X\bar{X}$: The sum $\sum\sum x_{ik}\bar{x}_k y_{ik} = \sum \bar{x}_k \sum x_{ik} y_{ik}$ requires the joint distribution of X and Y within each group, but these distributions may not always be available. The sum can be written

$$\sum \bar{x}_k(n_k r_k s_{xk} s_{yk} + n_k \bar{x}_k \bar{y}_k)$$

where r_k is the correlation coefficient between X and Y within the kth group. It may well be that this coefficient does not vary much from group to group. In that case the sum becomes

$$r_w \sum n_k \bar{x}_k s_{xk} s_{yk} + \sum n_k \bar{x}_k^2 \bar{y}_k$$

where r_w is the average within-group correlation coefficient.

If survey data are available for some or all the groups, a good estimate of r_w could be obtained from such data. In that case,

$$r_{(x\bar{x})y} = \frac{(r_w \sum n_k \bar{x}_k s_{xk} s_{yk} + (\sum n_k \bar{x}_k^2 \bar{y}_k)/n) - (\overline{x\bar{x}})\bar{y}}{s_{xx} s_y} \tag{I12}$$

X and $\bar{X}$: $\sum\sum x_{ik}\bar{x}_k = \sum n_k \bar{x}_k^2$

$$r_{x\bar{x}} = \frac{(\sum n_k \bar{x}_k^2/n) - \bar{x}^2}{s_x s_{\bar{x}}} \tag{I13}$$

X and $X\bar{X}$: $\sum\sum x_{ik} x_{ik} \bar{x}_k = \sum \bar{x}_k \sum x_{ik}^2 = \sum \bar{x}_k (n_k s_{xk}^2 + n_k \bar{x}_k^2)$

$$r_{x(x\bar{x})} = \frac{(\sum n_k \bar{x}_k s_{xk}^2 + (\sum n_k \bar{x}_k^2)/n) - \bar{x}(\overline{x\bar{x}})}{s_x s_{x\bar{x}}} \tag{I14}$$

$\bar{X}$ and $X\bar{X}$: $\sum\sum \bar{x}_k x_{ik} \bar{x}_k = \sum n_k \bar{x}_k^3$

$$r_{\bar{x}(x\bar{x})} = \frac{(\sum n_k \bar{x}_k^3/n) - \bar{x}(\overline{x\bar{x}})}{S_{\bar{x}} S_{x\bar{x}}} \tag{I15}$$

References

Alexander, K., and B. K. Eckland
1975 "Contextual effects in the high school attainment process." American Sociological Review 40(3):402–16.
Alissi, A. S.
1965 "Social influences on group values." Social Work 10(1):14–22.
Alker, H. R., Jr.
1969 "A typology of ecological fallacies." Pp. 69–86 in M. Dogan and S. Rokkan (eds.), Quantitative Ecological Analysis in the Social Sciences. Cambridge: MIT Press.
Allardt, E.
1968 "The merger of American and European traditions of sociological research: context analysis." Social Sciences Information 7(1):151–68.
Alwin, D. F.
1975 "The decomposition of effects in path analysis." American Sociological Review 40(1):37–40.
1976 "Assessing school effects: some identities." Sociology of Education 49(4): 294–303.
Asher, H.
1974 Causal Modeling. Quantitative Applications in the Social Sciences. Sage University Papers.
Bachman, J. G., C. G. Smith, and J. A. Slesinger
1966 "Control, performance, and satisfaction: an analysis of structural and individual effects." Journal of Personality and Social Psychology 4(2):127–36.
Barton, A. H.
1968 "Bringing society back in: survey research and macro-methodology." American Behavioral Scientist 12(2):1–9.
1970 "Allen Barton comments on Hauser's 'context and consex'." American Journal of Sociology 76(3):514–17.
Becker, G. S.
1972 "Schooling and inequality from generation to generation—Comment on a paper by Samuel Bowles." Journal of Political Economy 80(S):252.
Berelson, B. R., P. F. Lazarsfeld, and W. N. McPhee
1954 Voting: A Study of Opinion Formation in a Presidential Campaign. Chicago: University of Chicago Press (pp. 98–117).
Biddle, B. J., and E. J. Thomas
1966 Role Theory: Concepts and Research. New York: Wiley.
Blalock, H. M., Jr.
1961 "Changes in nonexperimental designs." Pp. 95–126 in Causal Inferences in Nonexperimental Research. Chapel Hill: University of North Carolina Press.
1963 "Correlated independent variables: the problem of multicollinearity." Social Forces 42(2):233–37.
1967 "Introduction." Pp. 1–37 in Toward a Theory of Minority-Group Relations. New York: Wiley.
1969 Theory Construction: From Verbal to Mathematical Formulations Englewood Cliffs, N.J.: Prentice-Hall.

Blau, P. M.
1957 "Formal organization: dimensions of analysis." American Journal of Sociology 63(1):58–69.
1960 "Structural effects." American Sociological Review, 25(2):178–93.
Boudon, R.
1969 "Secondary analysis and survey research: an essay in the sociology of the social sciences." Social Science Information 8(6):7–32.
Bowers, W.
1968 "Normative constraints on deviant behavior in the college context." Sociometry 31(4):370–85.
Bowles, S. S., and H. M. Levin
1968 "More on multicollinearity and the effectiveness of schools." Journal of Human Resources 3(3):393–400.
Boyd, L. H.
1971 Multiple Level Analysis with Complete and Incomplete Data. Ph.D. Dissertation, Political Science Department, University of Michigan microfilms, Ann Arbor.
Boyle, R. P.
1966a "The effect of the high school on students' aspirations." American Journal of Sociology 71(6):628–39.
1966b "On neighborhood context and college plans (III)." American Sociological Review 31(5):706–07.
Butler, D. E., and D. Stokes
1969 "Patterns in change." Pp. 293–312 in Political Change in Britain. New York: St. Martin's Press.
Cain, G. G., and H. W. Watts
1972 "Problems in making policy inferences from the Coleman report." Pp. 73–95 in P. H. Rossi and W. Williams (eds.), Evaluating Social Programs. New York: Seminar Press.
Campbell, E. G., and C. N. Alexander
1965 "Structural effects and interpersonal relationships." American Journal of Sociology 71(3):284–89.
Cartwright, D.
1969 "Ecological variables." Pp. 155–218 in E. F. Borgatta (ed.), Sociological Methodology. San Francisco: Jossey-Bass.
Cartwright, D., and A. Zander
1968 Group Dynamics: Research and Theory, 3rd ed. New York: Harper and Row.
Cattell, R. B., D. R. Saunders, and G. F. Stice
1953 "The dimensions of syntality in small groups." Human Relations 6(4): 331–56.
Cohen, J.
1968 "Multiple regression as a general data-analytic system." Psychological Bulletin 70(6):426–43.
Coleman, J. S.
1958–9 "Relational analysis: the study of social organizations with survey

methods." Human Organization 17(4):28–36.
1970 "Introduction." Pp. 3–4 in J. S. Coleman, A. Etzioni and J. Porter (eds.), Macrosociology: Research and Theory. Boston: Allyn and Bacon.

Collins, B. E., and H. Guetzkow
1964 A Social Psychology of Group Process for Decision-Making. New York: Wiley.

Converse, P. E., A. Campbell, W. E. Miller, and D. E. Stokes
1966 "Stability and change in 1960: a reinstating election." Pp. 78–95 in Elections and the Political Order. New York: Wiley.

Converse, P. E.
1969 "Survey research and the decoding of patterns in ecological data." Pp. 459–85 in M. Dogan and S. Rokkan (eds.), Quantitative Ecological Analysis in the Social Sciences. Cambridge: MIT Press.

Cowart, A.
1974 "A cautionary note on aggregate indicators of split ticket voting." Political Methodology 1(1):109–30.

Cox, K. R.
1969 "The spatial structuring of information flow and partisan attitudes." Pp. 157–85 in M. Dogan and S. Rokkan (eds.), Quantitative Ecological Analysis in the Social Sciences. Cambridge: MIT Press.

Cronbach, L. J., and Webb, N.
1975 "Between-class and within-class effects in a reported aptitude × treatment interaction: reanalysis of a study by G. L. Anderson." Journal of Educational Psychology 67(6):714–24.

Darlington, R. B.
1968 "Multiple regression in psychological research and practice." Psychological Bulletin 69(3):161–82.

Davies, J. A.
1959 "A formal interpretation of the theory of relative deprivation." Sociometry 22(4):280–96.
1961a Great Books and Small Groups. New York: The Free Press of Glencoe.
1961b "Compositional effects, role systems and the survival of small discussion groups." Public Opinion Quarterly 25(4):574–84.

Davis, J. A., J. L. Spaeth and C. Huson
1961 "A technique for analyzing the effects of group composition." American Sociological Review 26(2): 215–25.

Dogan, M., and S. Rokkan (eds.)
1969 Quantitative Ecological Analysis in the Social Sciences. Cambridge: MIT Press.

Duncan, O. D., and B. Davis
1953 "An alternative to ecological correlation." American Sociological Review 18(6):665–66.

Duncan, O. D., R. P. Cuzzort, and B. Duncan
1961 "Analysis of areal data." Pp. 60–174 in Statistical Geography: Problems in Analyzing Areal Data. Glencoe, Ill.: Free Press.

Duncan, O. D.
1970 "Partials, partitions, and paths." Pp. 38–47 in E. F. Borgatta and G. W. Bohrnstedt (eds.), Sociological Methodology. San Francisco: Jossey-Bass.
Duncan, O. D., D. L. Featherman, and B. Duncan
1972 Socioeconomic Background and Achievement. New York: Seminar Press.
Durkheim, E.
1951 Suicide: A Study in Sociology. Tr. J. A. Spaulding and G. Simpson. Glencoe, Ill.: Free Press.
Ennis, P. H.
1962 "The contextual dimension in voting." Pp. 180–211 in W. McPhee and W. Glaser (eds.), Public Opinion and Congressional Elections. Glencoe, Ill.: Free Press.
Eulau, H. (ed.)
1969 Micro-Macro Political Analysis: Accents of Inquiry. Chicago: Aldine Publishing.
Faris, R. E. L., and H. W. Dunham
1939 Mental Disorders in Urban Areas. Chicago: University of Chicago Press.
Farkas, G.
1974 "Specification, residuals, and contextual effects." Sociological Methods and Research 2(3):333–63.
Fennessey, J.
1968 "The general linear model: a new perspective on some familiar topics." American Journal of Sociology 74(1):1–27.
Festinger, L., S. Schachter, and K. W. Back
1950 Social Pressures in Informal Groups. New York: Harper and Brothers.
Firebaugh, G.
1977 "Assessing group effects: a comparison of two methods." Unpublished Manuscript. Dept. of Sociology and Anthropology. Vanderbilt University, Nashville.
1978 "A rule for inferring individual-level relationships from aggregate data." American Sociological Review 43(4):557–72.
Flinn, W. L.
1970 "Influence of community values on innovativeness." American Journal of Sociology 75(6):983–91.
Glasser, P., R. Sarri, and R. Vinter (eds.)
1974 Individual Change Through Small Groups. New York: Free Press.
Goldberger, A. S.
1964 Econometric Theory. New York: Wiley.
Goodman, L. A.
1953 "Ecological regressions and the behavior of individuals." American Sociological Review 18(6):663–64.
1959 "Some alternatives to ecological correlation." American Journal of Sociology 64(6):610–25.
Gordon, R.
1968 "Issues in multiple regression." American Journal of Sociology 73(5): 592–616.

Green, H. A. J.
1964 Aggregation in Economic Analysis. Princeton, N.J.: Princeton University Press.

Grunfeld, Y., and Z. Griliches
1960 "Is aggregation necessarily bad?" Review of Economics and Statistics 42(1):1–13.

Guetzkow, H.
1968 "Differentiation of roles in task oriented groups." Pp. 512–26 in D. Cartwright and A. Zander (eds.), Group Dynamics: Research and Theory. New York: Harper and Row.

Hammond, J. L.
1973 "Two sources of error in ecological correlations." American Sociological Review 38(6):764–77.

Hannan, M. T.
1971a Aggregation and Disaggregation in Sociology. Lexington, Mass.: Lexington Books.
1971b "Problems of aggregation." Pp. 473–508 in H. M. Blalock, Jr. (ed.), Causal Models in the Social Sciences. Chicago: Aldine, Atherton.

Hannan, M. T., and L. Burstein
1974 "Estimation from grouped observations." American Sociological Review 39(3):374–92.

Hanushek, E. A.
1974 "Efficient estimators for regressing regression coefficients." The American Statistician 28(2):66–7.

Hanushek, E. A., J. E. Jackson, and J. F. Kain
1974 "Model specification, use of aggregate data, and the ecological correlation fallacy." Political Methodology 1(1):640–56.

Hardor, T., and F. U. Pappi
1969 "Multiple-level regression analysis of survey and ecological data." Social Science Information 8(5):43–67.

Hare, A. P.
1962 Handbook of Small Group Research. New York: Free Press.

Hare, A. P., E. F. Borgatta, and R. F. Bales
1965 Small Groups: Studies in Social Interaction. New York: Alfred A. Knopf.

Hauser, R. M.
1968 Family, School, and Neighborhood Factors in Educational Performances in a Metropolitan School System. Ph.D. Dissertation, The University of Michigan microfilms, Ann Arbor.
1969 "Schools and the stratification process." American Journal of Sociology 74(6):587–611.
1970a "Context and consex." American Journal of Sociology 75(4):645–664.
1970b "Reply to Barton." American Journal of Sociology 76(3):517–20.
1971 Socioeconomic Background and Educational Performance. Washington, D. C.: American Sociological Association.
1974 "Contextual analysis revisited." Sociological Methods and Research 2(3): 365–75.

Hauser, R. M., W. H. Sewell, and D. F. Alwin
1976 "High school effects on achievement." in W. H. Sewell, R. M. Hauser and D. L. Featherman (eds.), Schooling and Achievement in American Society. New York: Academic Press.
Hoffman, P. J.
1962 "Assessment of the independent contributions of predictors." Psychological Bulletin 59:77–80.
Homans, C. H.
1964 "Bringing men back in." American Sociological Review 29(5):809–18.
Hyman, H. H., and E. Singer (eds.)
1968 Readings in Reference Group Theory and Research. New York: Free Press.
Irwin, G. A., and D. A. Meeter
1969 "Building voter transition models from aggregate data." Mideast Journal of Political Science 13(4):545–66.
Iversen, G. R.
1969 Estimation of Cell Entries in Contingency Tables When Only Margins Are Observed. Ph.D. Dissertation, Department of Statistics, Harvard University.
1973 "Recovering individual data in the presence of group and individual effects." American Journal of Sociology 79(2):420–34.
Kandel, D. B., and G. S. Lesser
1969 "Parental and peer influences on educational plans of adolescents." American Sociological Review 34(2):213–23.
Kendall, P. L., and P. F. Lazarsfeld
1950 "Problems of survey analysis." Pp. 133–96 in R. K. Merton and P. F. Lazarsfeld (eds.), Continuities in Social Research: Studies in the Scope and Method of 'The American Soldier'. Glencoe, Ill.: Free Press.
Kish, L.
1962 "A measurement of homogeneity in areal units." Bulletin of the International Statistical Association 4:201–10.
Kleinbaum, D. G., and L. L. Kupper
1978 Applied Regression Analysis and Other Multivariable Methods. North Scituate, Mass.: Duxbury Press.
Kousser, J.
1973 "Ecological regression and the analysis of past politics." Journal of Interdisciplinary History 4(2):237–62.
Lazarsfeld, P. F.
1959 "Problems in methodology." Pp. 39–78 in R. Merton (ed.), Sociology Today. New York: Basic Books.
Lazarsfeld, P. F., and H. Menzel
1961 "On the relation between individual and collective properties." Pp. 422–40 in A. Etzioni (ed.), Complex Organizations: A Sociological Reader. New York: Holt, Rinehart, and Winston.
Lee, T. C., G. G. Judge, and A. Zellern
1967 "Maximum likelihood and Bayesian estimation of transition probabilities."

Report 6733, Center for Mathematical Studies in Business and Economics, University of Chicago.

Lindzey, G., and E. Aronson (eds.)
1969 The Handbook of Social Psychology, 2nd ed. Reading, Mass.: Addison-Wesley.

Linz, J. J.
1969 "Ecological analysis and survey research." Pp. 91–131 in M. Dogan and S. Rokkan (eds.), Quantitative Ecological Analysis in the Social Sciences. Cambridge: MIT Press.

Lipset, S. M., M. A. Trow, and J. S. Coleman
1956 "Methodological note." Pp. 419–427 in Union Democracy. Glencoe, Ill.: Free Press.

Lott, A. J., and B. E. Lott
1961 "Group cohesiveness, communication level, and conformity." Journal of Abnormal and Social Psychology 62(2):408–12.
1965 "Group cohesiveness as interpersonal attraction: a review of relationships with antecedent and consequent variables." Psychological Bulletin 64(4): 259–309.

Mannheim, B. F.
1966 "Reference groups, membership groups and the self-image." Sociometry 29(3):265–79.

McDill, E. L., E. D. Meyers, Jr., and L. C. Rigsby
1967 "Institutional effects on the academic behavior of high school students." Sociology of Education 40(3):181–99.

McKeachie, B. F.
1954 "Individual conformity to attitudes of classroom groups." Journal of Abnormal and Social Psychology 49(2):282–89.

Meckstroth, T. W.
1970 "Measuring subsystem change with ecological data." Presented at the American Political Science Association Annual Meeting, Sept. 1970, Los Angeles, Calif.
1974 "Some problems in cross-level inference." American Journal of Political Science 18(1):45–66.

Menzel, H.
1950 "Comment." American Sociological Review 15(5):674.

Merton, R. K.
1957 "Continuities in the theory of reference groups and social structure." Pp. 281–386 in R. Merton (ed), Social Theory and Social Structure. Glencoe, Ill.: Free Press.

Meyer, J. W.
1970 "High school effects on college intentions." American Journal of Sociology 76(1):59–70.

Michael, J. A.
1961 "High school climates and plans for entering college." Public Opinion Quarterly 25(4):585–95.

1966 "On neighborhood context and college plans (II)." American Sociological Review 31(5):702–06.

Miller, J. L., and M. L. Erickson

1974 "On dummy variable regression analysis: a description and illustration of the method." Sociological Methods and Research 2(4):409–30.

Miller, W. E.

1956 "One-party politics and the voter." American Political Science Review 50(3):707–25.

Mills, T. M.

1967 The Sociology of Small Groups. Englewood Cliffs, N.J.: Prentice Hall.

Mosteller, F.

1968 "Association and estimation in contingency tables." Journal of the American Statistical Association 63:3–28.

Mueller, C. W.

1974 "City effects on socioeconomic achievements: the case of large cities." American Sociological Review 39(5):652–57.

Nagel, E.

1961 The Structure of Science. Chap. 11 and 14. New York: Harcourt, Brace and World.

Namboodiri, N. J., L. F. Carter, and H. M. Blalock, Jr.

1975 Applied Multivariate Analysis and Experimental Designs. New York: McGraw-Hill.

Nelson, J. I.

1972 "High school context and college plans: the impact of social structure on aspirations." American Sociological Review 37(2):143–48.

Nerlove, N., and S. J. Press

1973 Univariate and Multivariate Log-Linear and Logistic Models. Santa Monica, Calif.: Rand.

Orcutt, G. H., H. G. Watts, and J. B. Edwards

1968 "Data aggregation and information loss." American Economic Review 58(4):773–87.

Price, D.

1968 "Micro- and macro-politics: notes on research strategy." Pp. 102–40 in O. Garceau (ed.), Political Research and Political Theory. Cambridge, Mass.: Harvard University Press.

Przeworski, A.

1974 "Contextual models of political behavior." Political Methodology 1(1): 27–61.

Przeworski, A., and H. Teune

1970 The Logic of Comparative Inquiry. New York: Wiley.

Putnam, R. D.

1966 "Political attitudes and the local community." American Political Science Review 60(3):640–54.

Rapoport, A.

1963 "Mathematical models of social interaction." Pp. 493–580 in R. D. Luce,

R. R. Busch and E. Galenter (eds.), Handbook of Mathematical Psychology, Volume I. New York: Wiley.

Reiss, A. J., Jr., and A. L. Rhodes
1961 "The distribution of juvenile delinquency in the social class structure." American Sociological Review 26(5):720–32.

Rigsby, L. C., and E. L. McDill
1972 "Adolescent peer influence processes: conceptualization and measurement." Social Science Research 1:305–21.

Riley, M. W.
1964 "Sources and types of sociological data." Pp. 1015–21 in R. Farris (ed.), Handbook of Modern Sociology. Chicago: Rand McNally.

Robinson, W. S.
1950 "Ecological correlations and the behavior of individuals." American Sociological Review 15(3):351–57.

Rokkan, S.
1962 "The comparative study of political participation: notes toward a perspective on current research." Pp. 47–90 in A. Ranney (ed.), Essays on the Behavioral Study of Politics. Urbana, Ill.: University of Illinois Press.

Rosenberg, M.
1962 "The dissonant religious context and emotional disturbance." American Journal of Sociology, 68(1):1–10.

Scheuch, E. K.
1969 "Social context and individual behavior." Pp. 133–55 in M. Dogan and S. Rokkan (eds.), Quantitative Ecological Analysis in the Social Sciences. Cambridge: MIT Press.

Schuessler, K.
1969 "Covariance analysis in sociological research." Pp. 219–44 in E. F. Borgatta (ed.), Sociological Methodology. San Francisco: Jossey-Bass.
1971 Analyzing Social Data. Boston: Houghton-Mifflin.

Segal, D. K., and M. W. Meyer
1969 "The social context of political partisanship." Pp. 217–32 in M. Dogan and S. Rokkan (eds.), Quantitative Ecological Analysis in the Social Sciences. Cambridge: MIT Press.

Selvin, H. C.
1958 "Durkheim's 'Suicide' and problems of empirical research." American Journal of Sociology 63(6): 607–19.

Selvin, H. C., and W. O. Hagstrom
1963 "The empirical classification of formal groups." American Sociological Review 28(3):399–411.

Sewell, W. H.
1964 "Community of residence and college plans." American Sociological Review 29(1):24–38.

Sewell, W. H., and J. M. Armer
1966 "Neighborhood context and college plans." American Sociological Review 31(2):159–68.

Shively, W. P.
1969 "Ecological inference: the use of aggregate data to study individuals." American Political Science Review 63(4):1183–96.
1974 "Utilizing external evidence in cross-level inference." Political Methodology 1(4):61–73.
Sills, D. L.
1961 "Three 'Climate of Opinion' studies." Public Opinion Quarterly 25(4): 571–73.
Simpson, R. L.
1962 "Parental influence, anticipatory socialization and social mobility." American Sociological Review 27(4):517–22.
Slater, P. E.
1965 "Role differentiation in small groups." Pp. 498–15 in A. P. Hare, E. F. Borgatta and R. F. Bales (eds.), Small Groups: Studies in Social Interaction. New York: Alfred A. Knopf.
Slatin, G. T.
1969 "Ecological analysis of deliquency: aggregation effects." American Sociological Review 34(6):894–906.
Stokes, D. E.
1965 "A variance components model of political effects." Pp. 61–85 in Mathematical Applications in Political Science. Dallas: Arnold Foundation.
1966 "Analytical reduction in the study of institutions." Prepared for the 1966 Annual Convention of the American Political Science Association. Statler-Hilton Hotel, New York.
1969 "Ecological regression as a game with nature." Pp. 62–83 in J. L. Bernd (ed.), Mathematical Applications in Political Science, IV. Charlottesville: University Press of Virginia.
Stouffer, S. A., and others
1949 The American Soldier, Volume II. Princeton, N.J.: Princeton University Press.
Tannenbaum, A. S., and J. G. Bachman
1964 "Structural versus individual effects." American Journal of Sociology 69(6):585–95.
Tannenbaum, A. S., and C. G. Smith
1964 "The effects of member influence in an organization: phenomenology versus organization structure." Journal of Abnormal and Social Psychology 69(4):401–10.
Telser, L. G.
1963 "Least-squares estimates of transition probabilities." Pp. 270–92 in C. Christ et al. (eds.), Measurement in Economics. Stanford: Stanford University Press.
Thibaut, J. W., and H. H. Kelley
1959 The Social Psychology of Groups. New York: Wiley.
Thomas, E. J., and C. F. Fink
1963 "Effects of group size." Psychological Bulletin 60(4):371–84.

Turner, R. H.
1966 "On neighborhood context and college plans (I)." American Sociological Review 31(5):698–702.
Valkonen, T.
1969 "Secondary analysis of survey data with ecological variables." Social Science Information 8(6):33–36.
1969 "Individual and structural effects in ecological research." Pp. 53–68 in M. Dogan and S. Rokkan (eds.), Quantitative Ecological Analysis in the Social Sciences. Cambridge: MIT Press.
Verba, S.
1961 Small Groups and Political Behavior. Princeton, N.J.: Princeton University Press.
Walker, E. L., and R. W. Heyns
1962 An Anatomy for Conformity. Englewood Cliffs, N.J.: Prentice-Hall.
Ward, J. H., Jr.
1962 "Comments on 'the paramorphic representation of clinical judgment'." Psychological Bulletin, 59:74–76.
Werts, C. E., and R. L. Linn
1971 "Considerations when making inferences within the analysis of covariance model." Educational and Psychological Measurement 31(2):407–16.
Wilson, A. B.
1959 "Residential segregation of social classes and aspirations of high school boys." American Sociological Review 24(6):836–45.
Zander, A., and H. Medow
1965 "Strength of group and desire for attainable group aspirations." Journal of Personality 33(1):122–39.

Index

DATE DUE